HOSE

HOSE PRACTICES
SEVENTH EDITION

VALIDATED BY
THE INTERNATIONAL FIRE SERVICE TRAINING ASSOCIATION

PUBLISHED BY
FIRE PROTECTION PUBLICATIONS
OKLAHOMA STATE UNIVERSITY

COVER PHOTO COURTESY OF: JAY K. BRADISH

Dedication

This manual is dedicated to the members of that unselfish organization of men and women who hold devotion to duty above personal risk, who count on sincerity of service above personal comfort and convenience, who strive unceasingly to find better ways of protecting the lives, homes, and property of their fellow citizens from the ravages of fire and other disasters . . . **The Firefighters of All Nations**.

Dear Firefighter:

The International Fire Service Training Association (IFSTA) is an organization that exists for the purpose of serving firefighters' training needs. Fire Protection Publications is the publisher of IFSTA materials. Fire Protection Publications staff members participate in the National Fire Protection Association and the International Association of Fire Chiefs.

If you need additional information concerning our organization or assistance with manual orders, contact:

Customer Service, Fire Protection Publications, Oklahoma State University
930 North Willis, Stillwater, OK 74078-8045
1-800-654-4055   Fax: 405-744-8204

For assistance with training materials, recommended material for inclusion in a manual, or questions on manual content, contact:

Editorial Department, Fire Protection Publications, Oklahoma State University
930 North Willis, Stillwater, OK 74078-8045
405-707-3020   Fax: 405-744-8204   Email: editors@ifstafpp.okstate.edu

First Printing, March 1988
Second Printing, December 1989
Third Printing, March 2000

Oklahoma State University in compliance with Title VI of the Civil Rights Act of 1964 and Title IX of the Educational Amendments of 1972 (Higher Education Act) does not discriminate on the basis of race, color, national origin or sex in any of its policies, practices or procedures. This provision includes but is not limited to admissions, employment, financial aid and educational services.

Copyright © 1988 by the Board of Regents, Oklahoma State University

All Rights reserved.

ISBN 0-87939-075-1
Library of Congress 88-80199

Seventh Edition
Printed in the United States of America

Table of Contents

PREFACE viii
GLOSSARY x
INTRODUCTION 1
  Brief History 1
  Purpose and Scope 2

1   HOSE AND COUPLING CONSTRUCTION 5
  Fire Hose and Coupling Standards 5
    National Hose and Coupling Standards 6
  Hose Classification by Use 7
    Attack Hose 7
    Relay-Supply Hose 9
    Intake Hose 10
    Fire Extinguisher Hose 10
  Hose Classification by Construction 10
  How Fire Hose is Constructed 11
    Woven-Jacket Fire Hose Construction 12
    Rubber-Covered Hose Construction 15
    Braided Hose Construction 16
    Wrapped Hose Construction 16
  How Hose Couplings are Constructed 17
    Types of Couplings 17
  How Couplings are Attached to Hose 21
    Expansion Ring Method 21
    Screw-In Expander Method 21
    Collar Method 22
    Tension Ring Method 22
    Banding Method 22
  Review Questions 23

2   CARE, MAINTENANCE, AND TESTING 31
  Causes and Prevention of Hose Damage 31
    Mechanical Damage 31
    Heat and Cold Damage 34
    Mildew and Mold Damage 34
    Chemical Damage 35
  Causes and Prevention of Coupling Damage 36
    Coupling Repair 36
  General Care and Maintenance of Hose 37
    Washing, Drying, and Storage 37
    Inspection 40
  How to Recouple Fire Hose 40
    Couplings Attached with Expansion Rings 41
    Couplings Attached with Screw-In Expanders 44
    Couplings Attached with Bolted-On Collars 46
    Couplings Attached with Tension Rings 47

  How to Service Test Hose 48
    Test Site Preparation 48
    Safety at the Hose Testing Site 49
    Service Test Procedure 50
    Testing Unlined Linen Hose 52
    Vacuum Testing Hard Suction Hose 53
  Patching Hose 55
  Records of Hose Tests and Inspections 55
  Review Questions 57

3   HOSE APPLIANCES AND TOOLS 67

  Nozzles 67
    Solid Stream Nozzles 67
    Fog Nozzles 68
    Exposure Nozzles 68
    Applicator Nozzles 69
    Master Stream Devices 69
  Valves 71
    Gate Valves 71
    Ball Valves 71
    Butterfly Valves 71
    Floating Valves 72
    Clapper Valves 72
    Piston Valves 73
  Valve Devices 73
    Four-Way Hydrant Valve 73
    Automatic Hydrant Valves 74
    Manifolds and Water Thieves 75
    Wyes and Siameses 76
    In-line Relay Valves 76
    Intake Relief Valves 76
  Fittings 77
  Proportioners and Eductors 78
  General Care and Maintenance of Appliances 79
  Tools and Other Devices 79
    Spanner and Hydrant Wrenches 79
    Hose Straps, Ropes, and Chains 80
    Hose Control Device 80
    Hose Rollers 80
    Hose Jackets 81
    Hose Clamps 81
    Suction Hose Strainers 82
    Hose Bridges 82
    Hose Wringers 83
  Review Questions 84

4   BASIC METHODS OF HANDLING HOSE 93

  Making and Breaking Hose Connections 93

Connecting Hose — One-Person Methods 94
Connecting Hose — Two-Person Methods 95
Breaking a Tight Screw-Thread Connection 96
Connecting Hose to Fixed Fittings 98
Attaching a Nozzle to Hose 98
Hose Rolls 99
Straight Roll 99
Donut Roll 101
Hose Carries and Drags 108
Hose Carries 108
Hose Drags 119
Advancing Hoselines 125
Advancing Hose Up a Stairway 127
Advancing Hose Up a Ladder 128
Advancing a Booster Line 129
Operating Hoselines 130
One-Person Methods 131
Two-Person Methods 132
Three-Person Methods 134
Rolling a Loop to Advance a Charged Hoseline 135
Review Questions 136

5   SUPPLY HOSE LOADS AND LAYOUT PROCEDURES 143
Determining Which Hose Load to Use 143
Direction of Hose Lays 145
Hose Loading Guidelines 147
Basic Hose Loads 148
Accordion Load 148
Horseshoe Load 152
Flat Load 154
The Reel Load 157
Hose Lay Procedures 159
Determining Which Hydrant to Use 160
Making a Hydrant 160
The Forward Lay 162
The LDH Supply Lay 163
The Reverse Lay 165
The Split Lay 166
Review Questions 167

6   ATTACK HOSE LOADS AND LAYOUT PROCEDURES 175
Laying Attack Hose from the Hose Bed 175
Laying a Single Attack Hose in a Forward Lay 176
Laying a Single Attack Hose in a Reverse Lay 178
Hose Load Finishes 180
Finish for a Forward Lay 180
Finishes for Reverse Lays 181
Preconnected Hose Loads 189

	Reverse Horseshoe Loads	191
	Flat Load	198
	Triple Layer Load	200
	Minuteman Load	202
	Wyed Flat Load	204
Hose Packs		207
	Standpipe Pack	207
	Wildland Pack	209
Review Questions		216

7 SPECIAL HOSE OPERATIONS ... 223
Connecting Hard Suction Hose ... 223
Connecting Soft Sleeve Hose ... 223
Connecting Hose to a Portable Monitor ... 224
Kinking Hose to Shut Down a Charged Line ... 225
Retrieving a Charged "Wild Line" ... 226
Securing a Hoseline to a Ground Ladder ... 227
Hoisting a Hoseline ... 228
Passing a Hoseline Upward ... 229
Review Questions ... 231

REVIEW ANSWERS ... 233
INDEX ... 239

List of Tables

G.1	Hose Test Pressure Table	xv
1.1	Threads Per Inch/Millimeters	6
1.2	Performance Requirements for Hose	11
1.3	Use Classification/Type of Construction	12
1.4	Coupling Size Available for Hose Sizes	19
5.1	Amount of Water Delivered by Supply Hose	162

THE INTERNATIONAL FIRE SERVICE TRAINING ASSOCIATION

The International Fire Service Training Association is an educational alliance organized to develop training material for the fire service. The annual meeting of its membership consists of a workshop conference which has several objectives —

> . . . to develop training material for publication
> . . . to validate training material for publication
> . . . to check proposed rough drafts for errors
> . . . to add new techniques and developments
> . . . to delete obsolete and outmoded methods
> . . . to upgrade the fire service through training

This training association was formed in November 1934, when the Western Actuarial Bureau sponsored a conference in Kansas City, Missouri, to determine how all agencies that were interested in publishing fire service training material could coordinate their efforts. Four states were represented at this conference and it was decided that, since the representatives from Oklahoma had done some pioneering in fire training manual development, other interested states should join forces with them. This merger made it possible to develop nationally recognized training material which was broader in scope than material published by an individual state agency. This merger further made possible a reduction in publication costs, since it enabled each state to benefit from the economy of relatively large printing orders. These savings would not be possible if each individual state developed and published its own training material.

From the original four states, the adoption list has grown to forty-four American States; six Canadian Provinces; the British Territory of Bermuda; the Australian State of Queensland; the International Civil Aviation Organization Training Centre in Beirut, Lebanon; the Department of National Defence of Canada; the Department of the Army of the United States; the Department of the Navy of the United States; the United States Air Force; the United States Bureau of Indian Affairs; The United States General Services Administration; and the National Aeronautics and Space Administration (NASA). Representatives from the various adopting agencies serve as a voluntary group of individuals who govern policies, recommend procedures, and validate material before it is published. Most of the representatives are members of other international fire protection organizations and this meeting brings together individuals from several related and allied fields, such as:

> . . . key fire department executives and drillmasters,
> . . . educators from colleges and universities,
> . . . representatives from governmental agencies,
> . . . delegates of firefighter associations and organizations, and
> . . . engineers from the fire insurance industry.

This unique feature provides a close relationship between the International Fire Service Training Association and other fire protection agencies, which helps to correlate the efforts of all concerned.

The publications of the International Fire Service Training Association are compatible with the National Fire Protection Association's Standard 1001, "Fire Fighter Professional Qualifications (1981)," and the International Association of Fire Fighters/International Association of Fire Chiefs "National Apprenticeship and Training Standards for the Fire Fighter." The standards are an effort to attain professional status through progressive training. The NFPA and IAFF/IAFC Standards were prepared in cooperation with the Joint Council of National Fire Service Organizations of which IFSTA is a member.

The International Fire Service Training Association meets each July at Oklahoma State University, Stillwater, Oklahoma. Fire Protection Publications at Oklahoma State University publishes all IFSTA training manuals and texts. This department is responsible to the executive board of the association. While most of the IFSTA training manuals can be used for self-instruction, they are best suited to group work under a qualified instructor.

Preface

Firefighters have witnessed a number of significant changes in their profession during the past 20 years. New technology has produced tools, equipment, and apparatus that are more cost effective and efficient. It follows, therefore, that along with these improvements have come changes in methodology.

This manual provides an overview of the changes that have taken place during the last decade with one of the most essential pieces of equipment used by firefighters throughout the world: fire hose. **Hose Practices** is written to provide the latest technical information about the ways modern fire hose is constructed and how it should be maintained to give years of dependable service. The manual describes, in detail, a number of proven methods for loading and laying out hose in the most efficient manner. This is of paramount importance, especially in view of the fact that today's fire departments are performing fireground tasks with fewer personnel than ever before.

As with any comprehensive training manual, a tremendous amount of work is required to bring up-to-date technical information and fireground-tested methods together into a format that is easy to read and adaptable to the greatest number of fire department operations. With grateful acknowledgement for the many hours of volunteer work that have been devoted to this manual, the following individuals are recognized:

VALIDATING COMMITTEE

Chairman	*Vice-Chairman*	*Secretary*
Elmo Anderson	Glenn Boughton	Charles Tice
West Memphis, AR	Broken Arrow, OK	Marietta, GA

Members

John Backen	John Hoglund	Greg Noll	Charlie Scott
Huntington Beach, CA	College Park, MD	Washington, D.C.	Lubbock, TX
Darryl Boles	George Hughes	Bill Morey	Bob Young
Topeka, KS	Dallas, TX	Roswell, NM	Plano, TX

Other Members

Ray Bailey	George Carney	Curtis Holter	Robert Porter
F. Daniel Belanger	Carl N. Clanton	John Horn	Nick Renihan
Paul S. Benton	James Chapman	Frank W. Hubble	Ralph Scantlin
Robert Bloom	James Doyle	J. R. Jones	Dale Scobee
Paul Boecker	Elsworth Greer	Sherrill Miller	James Simms
Gerald Brinkman	Eric Haussermann	Charles L. Page	Roger K. Sweet
Ken Burton	Gene Hendricks		

Special thanks to the following departments and personnel for staging and photographing pictures:

California Department of Forestry, Fire Station 42, Chico, California: Division Chief John Hawkins; Battalion Chief Dick Tiller; Fire Apparatus Engineer Rich Batchelder; and the personnel of Station 42.

Edmond (Oklahoma) Fire Department: Fire Chief Ron Lloyd; Assistant Fire Chief David Barnes; Captain John Gibbon; and members of Engine 1, Green shift.

Paradise (California) Fire Department: Fire Chief Richard Landrum; Assistant Chief Jim Broshears; and members of "B" shift.

Plano (Texas) Fire Department: Fire Chief William Peterson; Division Chief George Caldwell; Captain Roger Smith; Firefighter (photographer) John Bucklew; and "B" shift members of Engine 6 and Medic 6. They assisted with many of the photographs that appear in Chapters 1-3.

Tulsa (Oklahoma) Fire Department Training Division: Deputy Drillmaster Fred Cotton; Assistant Drillmaster Kenneth Taylor; and members of Engine 4B and Ladder 2B.

Stillwater (Oklahoma) Fire Department: Fire Chief Jim Smith; Firefighters Bowdie Austin, Ed McManus, Mike Russell, Robert Kaale, and Randy Thompson.

Yukon (Oklahoma) Fire Department: Fire Chief Robert H. Noll; Assistant Chief Eddie P. Hogan; and personnel of the Green and Black shifts.

Columbia (Missouri) Fire Department.

Lower Providence Township, PA Volunteer Fire Company.

George Hughes of National Fire Hose Corporation for extra trips to Stillwater to supply technical assistance on hose manufacturing processes, care and maintenance, and hose laying procedures.

Special thanks to Bob Rose, Chico (California) Fire Department, for much work shooting and developing photography.

William Greenblatt, Photography, St. Louis, Missouri.

Jaffrey Fire Protection Company.

Jim Davies, Service Manager, Weis American Fire Equipment, Oklahoma City, Oklahoma.

Thanks also to Charles Donaldson of California, Maryland, a former staff member who made significant contributions during development of this manual.

And finally, our gratitude is extended to the following Fire Protection Publications staff members, whose time and talent made final publication of this manual possible:

David W. England, Senior Publications Editor
Lynne C. Murnane, Senior Publications Editor
Bill Westhoff, Senior Publications Editor
Carol Smith, Publications Specialist
Cindy Brakhage, Publications Specialist
Robert Fleischner, Publications Specialist
Suzanne Goodwin, Technical Writer
Don Davis, Coordinator, Publications Production
Ann Moffat, Graphic Designer
Laurie Zirkle, Graphic Artist
Mike McDonald, Graphic Artist
Karen Murphy, Phototypesetter Operator II/Metric Equivalencies
Desa Porter, Phototypesetter Operator II
Roger McKim, Research Technician
Scott Stookey, Research Technician
Pete Mauro, Research Technician

Gene P. Carlson
Editor

Glossary

A

ACCEPTANCE TESTING (PROOF TEST) — A test of coupled hose by the manufacturer at the request of the purchasing agency. The hose is subjected to extremely high pressures to ensure its ability to withstand the most extreme conditions in the field.

ACCORDION LOAD — An arrangement of fire hose in a hose bed or compartment in which the hose lies on edge with the folds adjacent to each other.

ADAPTER — A fitting for connecting hose couplings with dissimilar threads but with the same inside diameter.

ADJUSTABLE FLOW NOZZLE — A nozzle designed so that the amount of water flowing through the nozzle can be increased or decreased at the nozzle.

APPLIANCE — A device, other than a coupling, that is used with hose and through which water must pass.

APPLICATOR PIPE — A curved pipe attached to a nozzle for applying water precisely over a burning object.

ATTACK HOSE — Generally, any hose between the attack pumper and the nozzles to which it supplies water. Also, any hose used in a handline to control and extinguish fire. Minimum size 1½ inch (38 mm).

AUTOMATIC HYDRANT VALVE — A valve that, when connected to a hydrant, opens automatically to permit water to flow into the supply line; may be mechanically operated or radio controlled.

B

BALL VALVE — A valve having a ball-shaped internal component with a hole through its center that permits water to flow through when aligned with the waterway.

BANDING METHOD — A means of attaching a coupling to a fire hose with tightly wound strands of narrow-gauge wire or steel bands.

BLUNT START (HIGBEE CUT) — A flattened angle at the end of the thread that prevents cross-threading when couplings are connected.

BOOSTER HOSE — A reinforced, rubber-covered, rubber-lined hose generally carried on apparatus on a reel and used for extinguishment of incipient and smoldering fires.

BRAIDED HOSE — A nonwoven rubber hose manufactured by braiding one or more layers of yarn, each separated by a rubber layer, over a rubber tube and encased in a rubber cover.

BURST TEST — A destructive test on a 3 foot (1 m) length of hose to determine its maximum strength.

BUTTERFLY VALVE — A type of control valve that uses a flat baffle operated by a quarter-turn handle.

C

CALENDERING — A fire hose inner tube manufacturing process in which rubber is pressed between opposing rollers to produce a flat sheet. A tube is then formed by lapping and bonding together the edges of the sized sheet to form a tube.

CAST COUPLING — A coupling manufacturing process in which molten metal is poured into a mold, allowed to cool, then the mold is removed from the hardened coupling.

CHAIN HOSE TOOL — A tool used to carry, secure, and otherwise aid in handling hose.

CLAPPER VALVE — A hinged valve that permits the flow of water in one direction only.

COLLAR METHOD — A means of attaching a coupling with a two- or three-piece collar, which is bolted into place.

COMBINATION LAY — A hose lay in which two or more hoselines are laid in either direction — water source to fire or fire to water source.

CONTRACTUAL SLEEVE BINDING (CAS BINDING) — A method of attaching couplings with a tension ring, which compresses a nylon sleeve to lock the hose onto the coupling shank.

COUPLING — The fitting permanently attached to the end of a hose, used to connect two hoselines together or a hoseline to such devices as nozzles, appliances, discharge valves, or hydrants.

CURING — A manufacturing step in making fire hose; the process of applying heat and pressure to "set" the shape of the tube and to increase its smoothness.

D

DECK GUN (TURRET PIPE) — A large master stream appliance mounted on a pumper or trailer and connected directly to a pump.

DONUT ROLL — A length of hose rolled up for storage and transport.

DROP-FORGED COUPLING — A coupling made by raising and dropping a drop hammer onto a block of metal as it rests on a forging die, thus forming the metal into the desired shape.

DUTCHMAN — An extra fold placed along the length of a section of hose as it is loaded so that its coupling rests in proper position.

E

EDUCTOR — A portable proportioning device that injects a liquid, such as foam concentrate, into the water flowing through a hoseline.

EQUIPMENT STRIP — Removal of essential fire fighting tools and equipment at the fire scene before a pumper proceeds to the water source.

EXPANDER — A device that enlarges the expansion rings used for securing threaded couplings to fire hose. Also, the inner component of a screw-in expander coupling.

EXPANSION RING — A malleable metal band that binds fire hose to a threaded coupling by compressing the hose tightly against the inner surface of the coupling.

EXPANSION RING METHOD — A means of attaching a threaded coupling to a fire hose in which a metal expansion ring is placed inside the end of the hose, then expanded to compress the hose tightly against the inner surface of the coupling.

EXTINGUISHER HOSE — A braided, rubber-covered hose used on extinguishers, made to withstand pressures up to 1,250 psi (8 619 kPa).

EXTRUDED COUPLING — A coupling manufactured by the process of extrusion.

EXTRUSION — Shaping heated plastic or metal by forcing the molten material through dies.

F

FEMALE COUPLING — A threaded swivel device on a hose or appliance made to receive a male coupling of the same thread and diameter.

FILLER YARN (WEFT YARN) — The threads running crosswise in fabrics or woven hose.

FINISH — An arrangement of hose usually placed on top of a hose load and connected to the end of the load.

FIRE SERVICE HOSE — A specially constructed lined woven-jacketed hose designed to withstand the hazards of the fire scene.

FLAT LOAD — An arrangement of fire hose in a hose bed or compartment in which the hose lies flat with successive layers one upon the other.

FLOATING VALVE — A valve with a spring-loaded, dome-shaped disk within the waterway that is held in the closed position by both spring tension and internal water pressure. When incoming pressurized water flows against the disk from the outside, it opens to permit water to flow through the valve.

FORESTRY HOSE — A single-jacket, small-diameter hose used to combat fires in the forest and in other wildland settings.

FORGED COUPLING — A coupling formed by pounding a hot metal pellet into a forging die, which forms the metal into the desired shape.

FORWARD LAY — A method of laying hose from the water supply to the fire scene.

FOUR-WAY HYDRANT VALVE — A device that permits a pumper to boost the pressure in a supply line connected to a hydrant without interrupting the water flow.

FRONT BUMPER WELL — A hose compartment built into the front bumper of a pumper.

G

GATE VALVE — A control valve with a solid plate operated by a handle and screw mechanism. Rotating the handle moves the plate into or out of the waterway.

GATED WYE — A wye with manually operated valves to permit separate control of water to each line.

H

HARD SUCTION HOSE (HARD SLEEVE) — A noncollapsible, rubberized length of hose with a steel core that connects a pump to a source of water and is used for drafting.

HIGBEE CUT (BLUNT START) — A flattened angle at the end of the thread that prevents cross-threading when couplings are connected.

HORSESHOE LOAD — An arrangement of fire hose in a hose bed or compartment in which the hose lies on edge in the form of a horseshoe.

HOSE BED — The main hose-carrying area of a pumper or other piece of apparatus, designed for carrying hose.

HOSE BIN — A tray or compartment, often located on the running board or over a hose bed, for carrying extra hose.

HOSE BRIDGE (HOSE RAMP) — A device placed astride hose to prevent damage to hose from traffic passing over it.

HOSE CABINET — A recessed wall cabinet that contains a wall hydrant and preconnected fire hose for incipient fire fighting.

HOSE CAP — A threaded female fitting used to cap a hoseline or a pump outlet.

HOSE CARRY — A method of moving and deploying fire hose.

HOSE CLAMP — A mechanical or hydraulic device that compresses a fire hose to stop the flow of water.

HOSE CONTROL DEVICE — A device used to hold a charged hoseline in a stationary position for an extended period of time.

HOSE DRYER — An enclosed cabinet containing racks on which fire hose can be dried.

HOSE JACKET — The outer covering of a hose. Also, a device clamped over a hose to contain water at a rupture point or to join damaged or dissimilar couplings.

HOSE PACK — A compact bundle of hose, usually bound to facilitate moving.

HOSE RACK — A portable or fixed storage unit for fire hose.

HOSE RECORD — An individual history of a section of hose from the time it is purchased until it is taken out of service.

HOSE REEL — A cylindrical device upon which fire hose is manually or mechanically rolled for later deployment.

HOSE ROLLER — A tool for preventing damage to hose when it is dragged over a sharp surface such as the edge of a roof.

HOSE TEST GATE VALVE — A special valve designed to prevent injury caused by a burst hoseline during hose testing.

HOSE TOOL — A strap, rope, or chain with a handle suitable for placing over a ladder rung, used to carry and secure a hoseline.

HOSE TOWER — A structure from which fire hose can be hung to drain and dry.

HOSE WRINGER — A device used to remove water and air from large diameter hose.

HYDRANT WRENCH — A specially designed tool used to open and close a hydrant valve.

I

INCREASER COUPLING — An adapter fitting used to connect a hose or appliances of differing sizes.

INDUSTRIAL HOSE — Fire hose, usually of lighter construction than fire service hose, used by industrial fire brigades.

IN-LINE EDUCTOR — An eductor that is placed along the length of a hoseline.

IN-LINE RELAY VALVE — A valve placed along the length of a supply hose that permits a pumper to connect to the valve to boost pressure in the hose.

INTAKE HOSE — Hose used to connect a fire department pumper or a portable pump to a nearby water source. May be soft or hard suction hose.

INTAKE RELIEF VALVE — A valve designed to prevent damage to a pump from water hammer or any sudden pressure surge.

K

KINK — A sharp bend in a fire hose that restricts water flow.

KINK TEST — A test of hose under extreme conditions to ensure performance by folding the hose over on itself, securing it to maintain the kink, and pressurizing. Pressures vary with the type of hose.

L

LADDER PIPE — A master stream device mounted on the fly of an aerial ladder.

LDH (LARGE DIAMETER HOSE) — A relay-supply hose of 3½ to 6 inches (90 mm to 150 mm), used to move large volumes of water quickly with a minimum number of pumpers and personnel.

LINED HOSE — Fire hose composed of one or two woven outside jackets and an inside rubber lining.

LONGITUDINAL HOSE BED — A hose bed located to the side of the main hose bed, designed to carry preconnected attack hose.

M

MAKING A HYDRANT — The procedure for connecting to, and laying hose forward from, a fire hydrant.

MALE COUPLING — A hose nipple with protruding threads that fits into the thread of a female coupling of the same pitch and appropriate diameter and thread count.

MANIFOLD (PORTABLE HYDRANT) — A device that receives a supply of water and distributes it through valves to a number of hoses.

MASTER STREAM — Any of a variety of heavy, large-caliber water streams, usually supplied by siamesing two or more hoselines into a manifold device delivering 400 gpm (1 600 L/min) or more.

MEDIUM DIAMETER HOSE (MDH) — A hose of 2½ to 3 inches (65 mm to 77 mm) used for both fire fighting attack and for relay-supply purposes.

MONITOR — A portable master stream appliance consisting of a manifold, stream straightener, and nozzle.

MULTIPLE JACKET HOSE — A type of hose construction consisting of a combination of two separately woven jackets (double jackets), or two or more interwoven jackets, and lined with an inner rubber tube.

N

NATIONAL STANDARD THREAD — A screw thread of specific dimensions for fire service use, as specified in NFPA 1963, *Standard for Screw Threads and Gaskets for Fire Hose Connections.*

NOZZLE — An appliance on the discharge end of a hoseline that forms a fire stream of definite shape, volume, and direction.

P

PIERCING APPLICATOR NOZZLE — A nozzle with an angled, case-hardened steel tip that can be driven through a wall, roof, or ceiling to extinguish hidden fire.

PISTON VALVE — A valve with an internal piston that moves within a cylinder to control the flow of water through the valve.

PORTABLE HYDRANT (MANIFOLD) — A device that receives a supply of water and distributes it through valves to a number of hoses.

PRECONNECT — An attack hose connected to a discharge when the hose is loaded; done to shorten the time it takes to deploy the hose for fire fighting.

PROPORTIONERS — A fixed mixing device that injects an extinguishing agent, such as foam, into a water stream.

Q

QUARTER-TURN COUPLING — A sexless coupling with two hooklike lugs that slip over a ring of the opposite coupling, then rotate 90 degrees clockwise to lock.

R

REDUCER COUPLING — An adapter fitting used to connect hose or appliances of differing sizes.

REDUCING WYE — A wye that has two outlets smaller in diameter than the inlet valve.

REEL LOAD — An arrangement of fire hose, especially large diameter hose, on a reel.

RELAY-SUPPLY HOSE (SUPPLY HOSE) — The hose between the water source and the attack pumper, laid to provide large volumes of water at low pressure.

REVERSE LAY — A method of laying hose from the fire scene to the water supply.

S

SCREW-IN EXPANDER METHOD — Some types of hose, particularly rubber-jacket booster hose, having threaded couplings attached with expanders that are screwed into place.

SERVICE TEST — Hydrostatic pressure testing of fire hose by the fire department.

SEXLESS COUPLING — A coupling with no distinct male or female components.

SHELL — The outer component of a screw-in expander coupling.

SIAMESE — A hose appliance with two or more female inlets and one male outlet.

SINGLE-JACKET HOSE — A type of hose construction consisting of one woven jacket, usually lined with an inner rubber tube.

SMALL DIAMETER HOSE (SDH) — A hose of ¾ to 2 inches (20 mm to 50 mm) used for fire fighting purposes.

SNAP COUPLING — A coupling set with nonthreaded male and female components. When a connection is made, two spring-loaded hooks on the female coupling engage a raised ring around the shank of the male coupling.

SOFT SLEEVE (SOFT SUCTION) — A short length of large diameter fire hose used to connect a pumper to a fire hydrant.

SOLID STREAM — A hose stream that stays together as a solid mass, as opposed to a fog or spray stream.

SPLIT LAY — A hose lay laid by two pumpers, one making a forward lay and one making a reverse lay from the same point.

STANDPIPE HOSE — A single-jacket hose, lined or unlined, that is preconnected to a standpipe. Used primarily by building occupants to mount a quick attack on an incipient fire.

STORZ COUPLING — A sexless coupling commonly found on large diameter hose.

SUPPLY HOSE — The hose between the water source and the attack pumper, laid to provide large volumes of water at low pressure.

T

TENSION RING METHOD — A method used to attach a coupling to large diameter hose using a tension ring and contractual sleeve.

THERMOPLASTIC — A plastic that softens with an increase of temperature and hardens with a decrease of temperature, but does not undergo any chemical change.

THREADED COUPLING — A male or female coupling with a spiral thread.

THREE-PLY PROCESS — A process of producing rubber-covered hose in which a nitrile rubber is vulcanized to the interior surface of a woven polyester tube.

TIER — A layer of hose loaded in the hose bed of a fire apparatus.

TRANSVERSE HOSE BED — A hose bed that lies across the pumper body, at a right angle to the main hose bed; designed to deploy preconnected attack hose to the sides of the pumper.

TURRET PIPE (DECK GUN) — A large master stream appliance mounted on a pumper or trailer and connected directly to a pump.

U

UNLINED HOSE — Fire hose without a rubber lining, most frequently used in interior standpipe systems and in wildland fire fighting.

V

VALVE — A water control device with an internal component that can be moved within the waterway to regulate the flow of water.

W

WARP YARN — The threads that run lengthwise in a fabric or woven hose.

WATER CURTAIN — A fan-shaped stream of water applied between a fire and an exposed surface to prevent the surface from igniting from radiated heat.

WATER HAMMER — The noise and force that develops in water hoses or pipes when the flow is suddenly stopped by the rapid closing of a valve or faucet.

WATER THIEF — Variation of a gated wye that allows the use of several attack lines.

WATERWAY — The path through which water flows within a hose.

WEEPING — Coupling leakage at the point of attachment.

WEFT YARN (FILLER YARN) — The threads running crosswise in fabric or woven hose.

WILD LINE — An uncontrolled hoseline and nozzle or butt that thrashes about from the reaction of highly pressurized, flowing water.

WOVEN-JACKET HOSE — A fire hose constructed with one or two outer jackets woven on looms from cotton or synthetic fibers.

WRAPPED HOSE — A nonwoven rubber hose manufactured by wrapping rubber-impregnated woven fabric around a rubber tube and encasing it in a rubber cover.

WYE — A hose appliance with one female inlet and two or more male outlets, usually gated.

TABLE G.1
HOSE TEST PRESSURE TABLE

Attack Hose Minimum Requirements*

Burst Test Pressure	900 psi	6 250 kPa
Proof Test Pressure	600 psi	4 140 kPa
Kink Test Pressure	450 psi	3 100 kPa
Service Test Pressure	300 psi	2 070 kPa

Operating Pressure not to Exceed 90% of Service Test Pressure

Large Diameter Supply Hose Minimum Requirements*

Burst Test Pressure	600 psi	4 140 kPa
Proof Test Pressure	400 psi	2 760 kPa
Kink Test Pressure	300 psi	2 070 kPa
Service Test Pressure	200 psi	1 380 kPa

Operating Pressure not to Exceed 185 psi (1 275 kPa)

*Based on NFPA 1962, 1987 Edition

Introduction

As the twenty-first century rapidly approaches, the fire service continues to deal with a problem civilization has faced for thousands of years — the devastation of uncontrollable fire. Amazingly, there has been little change over the centuries in the basic way we fight fire. Firefighters worldwide still put fire out by moving water through fire hose and directing it onto the fire. Water continues to serve as the cheapest, most effective fire extinguishing agent and, of all the equipment carried on a fire department pumper, fire hose is the one item most frequently used for extinguishing fire. Fire hose, therefore, is absolutely essential to accomplishing our primary mission: saving life and preserving property from damage or total loss caused by fire. The well-coordinated and efficient deployment of hose significantly increases the effectiveness of the overall fire attack. The laying of hose quickly, with no delay, is often crucial to stopping the progress of fire before it reaches an uncontrollable state.

Interestingly, too, hose laying operations play a strong role in the public's perception of firefighters. An aggressive fire attack reflects a high level of tactical competency. Such an operation requires careful preparation of apparatus and hose and frequent training in the methods of swiftly placing it into service.

During the early days of fire fighting, the process of conveying water to a fire and applying it for fire extinguishment was a major problem. Although a crude type of fire hose was developed in the seventeenth century, satisfactory fire hose was not produced until the nineteenth century. Until that time, the only means of transporting water to a fire was by passing buckets of water down a line of organized firefighters. This system of moving water was known as the "Bucket Brigade" (Figure I.1) and these brigades were often governed by laws under the direction of Fire Masters.

Figure I.1 An early means of transporting and applying water to a fire was the "Bucket Brigade."

2 HOSE

An important advancement in fire fighting equipment was made when the hand-powered pump came into use. Flexible hose for this pump was made by sewing or riveting together the edges of long strips of leather. Although it leaked badly and was extremely heavy and unwieldy, this first hose was a major technological improvement over the bucket method of applying water. The next significant advance in hose manufacturing technology came with the introduction of circular looms, which were designed to form canvas tubes, woven from flax, that were much lighter and more maneuverable than leather. This type of hose was more leakproof than leather, but more vulnerable to damage from mildew and decay. The next principal improvement in fire hose came with the discovery of rubber. The woven tube was lined with an inner rubber tube to make the hose virtually leakproof and more resistant to bursting from internal pressure.

Since the introduction of rubber, many improvements have been made in the quality of materials used in the outer jacket and its inner lining. As a result, today's fire hose is lighter, more durable, and easier to handle than ever before. Along with this, the fire service has improved its methods of carrying and deploying hose from apparatus. While it is difficult to imagine how fire hose could become any better, the development of space-age materials will surely continue to improve the quality of this vital component of the fire fighting arsenal and ultimately reduce the tragic loss of lives and property from the ravages of fire.

PURPOSE AND SCOPE

The purpose of this manual is not only to fulfill the requirements of NFPA 1001, *Fire Fighter Professional Qualifications*, that relate to fire hose, but also to serve as a comprehensive single-source of information about fire hose and its use.

The information presented in this manual includes the construction methods and maintenance of fire hose, special uses and operations, and basic hose loads and handling techniques. No single technique is represented as the best practice; any one or several of the methods presented may be selected for use by a fire fighting agency. Descriptions of techniques have been kept general so that they may be adapted to local situations.

SPECIAL NOTE

One of the difficulties in writing a new edition of a manual about hose is that certain terms are used differently in different parts of the country, especially in relation to the way hose is stored, transported, and laid out. To eliminate confusion, this manual uses terms that most clearly identify the items or methods described. In some cases, this means that terms used in earlier editions of this manual are either changed or simply eliminated. We regret any inconvenience that results from these changes, but hope that this will ultimately simplify the instructional process.

1

Hose and Hose Coupling Construction

4 HOSE

This chapter provides information that addresses performance objectives described in NFPA 1001, *Fire Fighter Professional Qualifications* (1987), particularly those referenced in the following sections:

3-13 Fire Hose, Nozzles, and Appliances

3-13.1

Chapter 1
Hose and Coupling Construction

Fire hose is used to move water from one location to another. Because hose is used for a number of functions during fire fighting operations, there are many different types of hose. It is manufactured in different lengths and sizes; further, it is made of natural or synthetic materials, lined or unlined, and has different types and sizes of couplings.

Of all the hose manufactured for any purpose, *fire* hose must be manufactured of the highest quality according to the highest standards. Fire service hose, therefore, is of very durable construction because it is used frequently and must withstand the wear that occurs with daily use. Most fire service hose in use today is of the woven-jacket type of construction. There is, however, a movement toward using a rubberized hose made by a coating or extrusion process. Industrial hose is similar to fire service hose but may be of lighter construction because it is used less frequently than fire service hose and need not withstand the wear of daily use. Mill hose is single-jacket, lined hose used only for cleanup and washdown around mill yards and factories; it is not sold for fire protection use.

A thorough knowledge of hose is necessary to use it more effectively and to care for it properly. This knowledge should include the methods of hose and coupling construction, the types of hose and the respective uses for each type, as well as the limitations of each type of hose. This chapter begins with a short history of the standardization of hose and couplings. It goes on to describe the basic hose classifications, hose performance requirements, and hose construction. The last part of the chapter covers couplings — types, construction, and the methods by which couplings are attached to hose.

FIRE HOSE AND COUPLING STANDARDS

One of the great dilemmas facing the fire service at the turn of the century was that each fire department was using hose and fittings of different sizes and thread types. There were once over 2,000 different fire hose coupling threads in the United States. This problem was most apparent when fire departments were required to join forces during major fires and conflagrations. More often than not it was impossible for apparatus from one department to connect to hydrants and apparatus of other departments. This, of course, resulted in a greatly impaired fire fighting effort. As a result, needless lives and property were lost because water could not be efficiently moved to the fire.

During the late 1800s fire service administrators began to recognize the problems associated with each fire agency using couplings with different threads. In 1873, the newly organized International Association of Fire Engineers (now the International Association of Fire Chiefs) passed its first resolution — to set standards for fire hose screw threads. As early as 1898, the National Fire Protection Association (NFPA) began drawing up thread specifications for fire hose couplings. This led to the appointment of an NFPA committee in 1905 to standardize not only couplings, but also fire hose, nozzles, and accessories.

Ultimately, dimensions for 10 coupling sizes were developed. This process started in 1925, when dimensions for 2½-, 3-, 3½-, and 4½-inch (65 mm, 77 mm, 90 mm, 115 mm) couplings were agreed upon and adopted as an American national standard. In 1935, ¾-, 1-, and 1½-inch (20 mm, 25 mm, and 38 mm) thread dimensions were standardized, followed by 4-, 5-, and 6-inch (100 mm, 125 mm, and 150 mm) threads in 1955. In 1956, standard dimensions for the gasket grooves and gaskets for all 10 sizes were adopted. These standards were eventually adopted by such agencies as the American Insurance Association, the Association of Factory Mutual Insurance Companies, the International Association of Fire Chiefs, and the American Water Works Association.

The thread used for the 10 sizes of threaded fire hose couplings is specified in NFPA 1963, *Standard for Screw Threads and Gaskets for Fire Hose Connections*. This standard specifies the number of threads per inch for each size of coupling (Table 1.1), as well as lists the dimensions for machining and testing the accuracy of the threads.

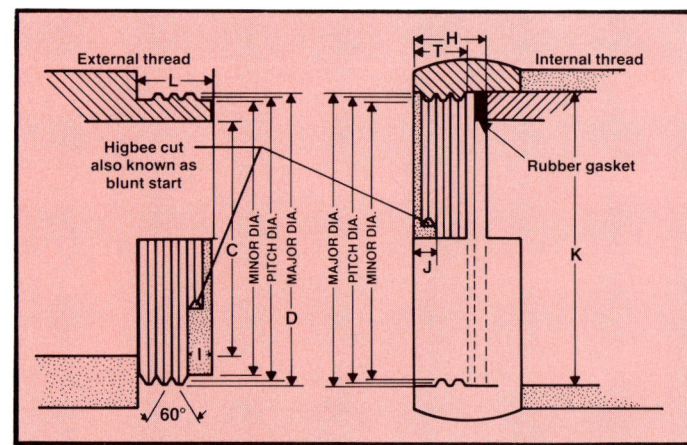

Figure 1.1 Specifications for a 2½-inch (65 mm) coupling with National Standard Threads. *Courtesy of National Fire Protection Association.*

**TABLE 1.1
THREADS PER INCH (MILLIMETERS)**

DIAMETER		NUMBER OF THREADS
INCHES	MM	
0.75	20	8
1.00	25	8
1.50	38	9
2.50	65	7½
3.00	77	6
3.50	90	6
4.00	100	4
4.50	115	4
5.00	125	4
6.00	150	4

The standardized thread is called the "American National Fire Hose Connection Screw Thread" (abbreviated "NH"). Because it is also commonly referred to as the "National Standard Thread," it is often abbreviated as "NST" or "NS". A detailed drawing of the National Standard Thread is shown in Figure 1.1.

National Hose and Coupling Standards

National standards for fire hose and couplings are now written by several organizations: the American National Standards Institute (ANSI); Underwriters Laboratories (UL); the Rubber Manufacturers Association (RMA); the American Society for Testing and Materials (ASTM); Factory Mutual (FM); the United States Department of Agriculture (USDA); and the United States Forest Service (USFS). The following is a listing of a number of current standards that apply to each of the four types of hose (designated by *A*-Attack; *S*-Relay-Supply; *I*-Intake; *E*-Extinguisher):

- NFPA 1961 *Standard for Fire Hose* (A,S,I)
- NFPA 1962 *Standard for the Care, Use, and Maintenance of Fire Hose, Including Connections and Nozzles* (A,S,I)
- NFPA 1963 *Standard for Screw Threads and Gaskets for Fire Hose Connections* (A,S,I)
- UL Standard 19 *Woven-Jacketed Rubber-Lined Fire Hose* (A,S,I)
- ASTM D296 *Woven-Jacketed Rubber-Lined Fire Hose for Public and Private Department Use* (A,S,I)
- FM 2111 *Approval Standard for 1½- and 2½-inch Lined Fire Hose* (A,S,I)
- USDA/USFS 5100-186a *Hose, Cotton-Synthetic Jacketed, Lined 1-inch and 1½-inch Hose* (A)
- USDA/USFS 5700-183f *Specification for 1- and 1½-inch Linen Hose* (A)
- NFPA 14 *Standard for Standpipe and Hose Systems* (A)

- NFPA 24 *Standard for Outside Protection* (A)
- ASTM D380 *Standard Method of Testing Rubber Hose* (A,I)
- ANSI A152.1 *Safety Standard for Unlined Fire Hose* (A)
- ANSI/UL 92 *Fire Extinguisher and Booster Hose* (A,E)
- RMA, IP-12 *Standard for High Pressure Fire Engine Booster and Fire Extinguisher Hose* (A,E)
- ASTM E380-78 *Standard for Metric Practice* (A,S,I,E)
- ANSI Z210.1 *Metric Practice Guide* (A,S,I,E)

HOSE CLASSIFICATION BY USE

The fire service describes hose in a number of ways. This manual places the many types of hose into four broad categories, based on the way the hose is used:

- Attack hose
- Relay-supply hose
- Intake hose
- Extinguisher hose

Attack Hose

This group of hose is by far the broadest because there are many types of attack hose on the market. For the sake of clarity, attack hose is defined as any hose that is used to directly control and extinguish fire. Attack hose can be subcategorized into the following groups:

Fire department hose is heavy-duty, lined hose carried for daily fire fighting use on fire apparatus (Figure 1.2). It may be of a fabric-jacket construction or of a rubber-covered construction. Woven fabric-jacket hose may be single-jacketed or multiple-jacketed. Most fabric-jacket fire department hose is double-jacketed and lined for maximum durability. Some newer, lightweight types of jacketed hose are made with a thermoplastic, rather than rubber, liner. Rubber-covered hose may be jacketed and lined or may be constructed

Figure 1.2 Fire department hose is constructed for heavy-duty use. *Courtesy of Phil Williamson.*

so that rubber and fabric are bonded into a single, inseparable unit. Both fabric-jacket and rubber-covered hoses are made in a variety of sizes. The criteria for attack hose is its service test pressure as listed on page xv.

Forestry hose is a single-jacket, small diameter hose used to combat fires in the forest and in other wildland settings (Figure 1.3). Made in both lined and unlined versions, some hoses are actually designed to leak water through the outer jack-

Figure 1.3 Forestry hose is relatively lightweight so that it can be carried over a variety of terrains. *Courtesy of John Fell.*

et. This "self-protecting" feature makes the hose more resistant to damage by heat when it is pulled through burning debris.

The small size and lightweight characteristic of forestry hose is necessitated by the requirement that it be carried long distances over uneven terrain, often uphill, and be used in a way that conserves water. In some regions it can be used with a pumper in a mobile fire attack. In less accessible terrain it can be backpacked or air-dropped to a remote scene and used with a portable pump that drafts water from a stream or other open water source. Forestry hose is also used with apparatus when wildland fires occur in areas that are accessible by road and have some sort of water supply.

Standpipe hose, either lined or unlined, is usually a single-jacket hose used primarily by building occupants to mount a quick attack on an incipient fire. The most common size is 1½ inch (38 mm), although other sizes are available. In some cases, standpipe hose is used by private fire brigades that lack fire apparatus and a standard fire hose complement. It is usually preconnected to a small diameter domestic standpipe or to a yard hydrant. Complete with a nozzle, it is suspended, accordion-like, from a rack within a cabinet (Figure 1.4) or stored on a reel.

A disadvantage of standpipe hose, particularly with the unlined type, is that it becomes unreliable with age. NFPA 1962, *Standard for the Care, Use, and Maintenance of Fire Hose, Including Connections and Nozzles*, advocates periodic inspection and testing of this hose. Testing the hose after several years of storage in the rack typically causes ruptures at the folds. When unlined standpipe hose fails service testing, it should be replaced with lined hose.

Because any type of domestic standpipe hose is typically unreliable, firefighters should only use fire department hose on standpipe systems. Lightweight, hose folded into a portable pack or bundle (Figure 1.5) can be carried on the apparatus. These "high-rise" hose packs can then be conveniently carried into the building and connected to the fire service standpipe system.

Figure 1.5 Lightweight hose folded into a portable "high-rise" pack can be carried on the apparatus. *Courtesy of Clemens Industries.*

Booster hose is a rubber-covered hose made of several layers of braided, rubberized material. Usually carried on reels (Figure 1.6), it is used to extinguish relatively small fires and is often used for overhaul work. Booster hose is manufactured in ¾- and 1-inch (20 mm and 25 mm) sizes. Because booster hose is of such a small diameter, it must be used at a high pressure to overcome the friction loss that occurs when producing the maximum water flow. Its construction is detailed later in this chapter.

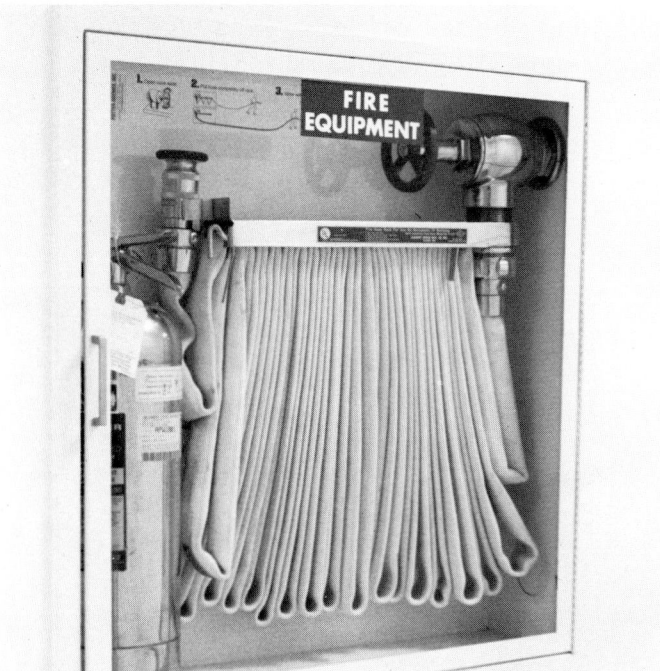

Figure 1.4 This standpipe hose is preconnected and accordion-folded on a rack.

Hose and Coupling Construction **9**

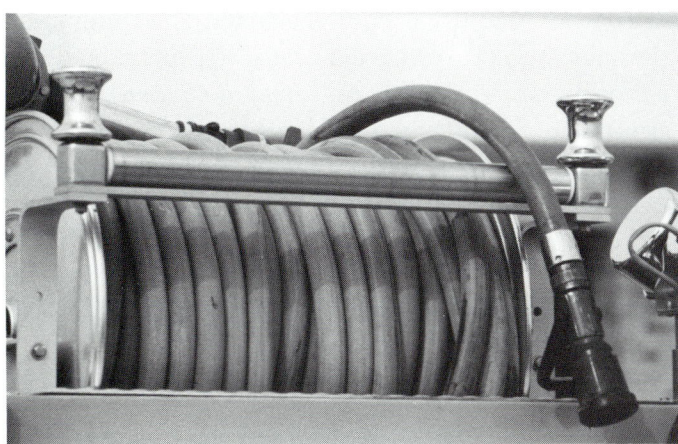

Figure 1.6 Booster hose is a braided, rubber-covered hose usually carried on reels.

Supply Hose

Supply hose is designed to move large volumes of water at low pressure. It is generally larger than attack hose. There are two groups within this category: medium diameter hose and large diameter hose.

Medium diameter hose (MDH) is a term given to supply hose of the 2½- or 3-inch (65 mm or 77 mm) diameter sizes. This hose, in effect, is attack hose used for both fire fighting and supply purposes. In a supply situation, MDH gives satisfactory service where operations do not routinely require large volumes of water and when layout distances are relatively short. The problem with using attack hose for long-distance water supply, however, is that the pressure loss caused by water friction on the inner lining of the hose seriously depletes the water flow. When large volumes of water must be delivered with medium diameter hose, several parallel lines must be laid (Figure 1.7) or pumpers must be placed at intervals along the hose lay to boost the pressure (Figure 1.8).

Large diameter hose (LDH) was developed to overcome the pressure loss problems of medium diameter hose. Available in both woven-jacket or rubber-covered versions, LDH sizes for relay and supply are 3½, 4, 4½, 5, and 6 inch (90 mm, 100 mm, 115 mm, 125 mm, and 150 mm). It is designed to move large volumes of water quickly with a minimum number of pumpers and personnel. It should be noted, however, that LDH can be used to move smaller volumes of water with virtually no loss of pressure.

Figure 1.7 When medium diameter hose is used in a high-volume supply situation, several hoses are required.

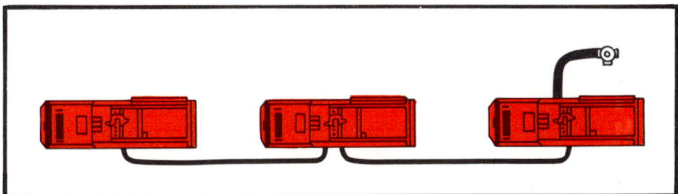

Figure 1.8 Pumpers set up in relay help boost the pressure in medium diameter hose when moving water over a long distance.

The largest sizes of LDH make it possible for a pumper to lay the hose from a hydrant to the fire scene without the need for a second pumper to boost the pressure at the hydrant (Figure 1.9).

Figure 1.9 Using large diameter hose often eliminates the need to have a pumper at the hydrant to boost the pressure. *Courtesy of Plano, Texas Fire Department.*

Intake Hose

Intake hose is used to connect a fire department pumper or a portable pump to a nearby water source. There are two groups within this category: soft sleeve hose and hard suction hose. Soft sleeve hose (also called a soft suction) is used to transfer water from a pressurized water source, such as a fire hydrant, to the pump intake (Figure 1.10). It is usually of a multiple-jacket, lined construction but lightweight versions are now appearing on the market. Soft sleeves are available in sizes ranging from 2½ to 6 inches (65 mm to 150 mm). Hard suction hose (also called a hard sleeve) is used primarily to draft water from an open water source (Figure 1.11). It is also used to siphon water from one portable tank to another, usually in connection with a tanker shuttle operation. Hard suction hose is constructed of a rubberized, reinforced material designed to withstand the partial vacuum conditions created when drafting. It is also available in sizes ranging from 2½ to 6 inches (65 mm to 150 mm).

Fire Extinguisher Hose

Fire extinguisher hose is used on large extinguisher units that may be stationary, wheeled (Figure 1.12), or vehicle mounted. The hose, which may be coiled or on reels, transports liquid, gaseous, or powder extinguishing agents from the extinguisher container to the nozzle. There are two groups of fire extinguisher hose: conventional extinguisher hose is used with extinguishers that discharge at pressures no greater than 400 psi (2 758 kPa); high-pressure extinguisher hose will withstand pressures of up to 1,250 psi (8 619 kPa). Both conventional and high-pressure hoses are made in essentially the same way as booster hose.

HOSE CLASSIFICATION BY CONSTRUCTION

Another way to classify hose is according to the way it is constructed. Hose construction standards specify a number of construction and performance characteristics. The standards require that all fire hose move water reliably and efficiently, be maneuverable, be durable, and resist kinking. The design and manufacture of hose varies according to the purpose for which it is used. Each use requires

Figure 1.10 A soft sleeve hose transfers water from the fire hydrant to the pump intake. *Courtesy of Plano, Texas Fire Department.*

Figure 1.11 Hard suction hose is designed to withstand the partial vacuum of drafting. *Courtesy of Plano, Texas Fire Department.*

Hose and Coupling Construction 11

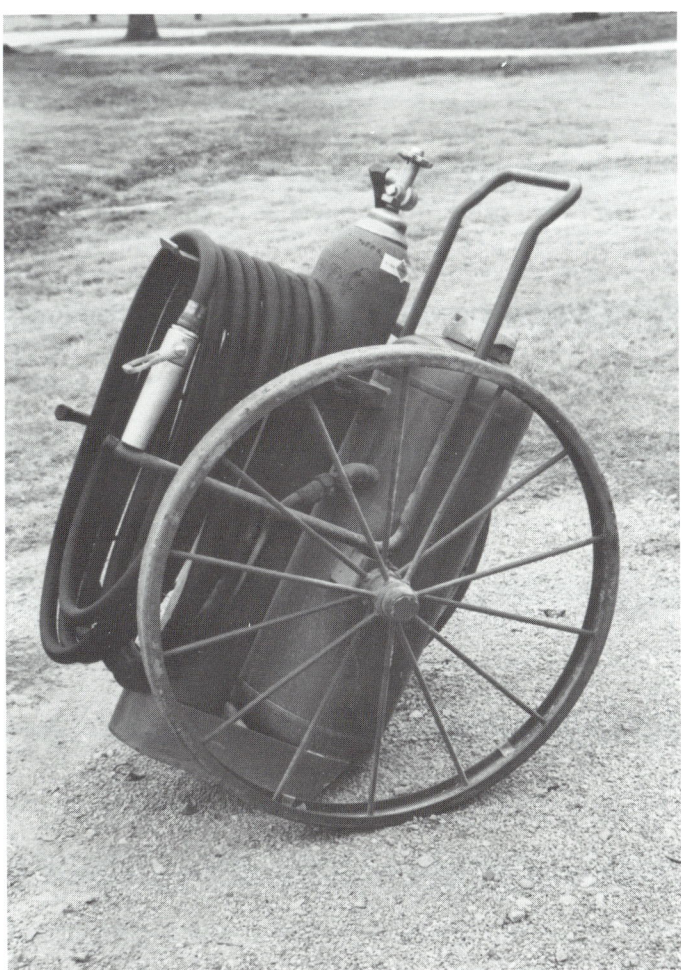

Figure 1.12 Fire extinguishers, such as this wheeled unit, must be equipped with specially made fire extinguisher hose.

TABLE 1.2
PERFORMANCE REQUIREMENTS FOR HOSE

USE CATEGORY	1	2	3	4	5	6	7	8	9	10	11	12	13	14	15
ATTACK															
Fire Dept.			●			●	●			●					●
Forestry				●	●	●	●						●	●	●
Booster	●	●					●	●					●		●
Standpipe				●	●	●	●					●			
RELAY-SUPPLY															
Medium Diameter							●				●				●
Large Diameter		●									●	●			●
INTAKE															
Soft Sleeve						●					●	●			
Hard Suction		●									●	●		●	
FIRE EXTINGUISHER															
Conventional	●	●				●		●				●	●		
High Pressure	●	●					●	●	●			●	●		

1. Semi-rigid or noncollapsible
2. Does not require drying
3. Extremely durable
4. Ultra-light weight
5. Extra compact for storage
6. Highly flexible
7. Easily moved by one person
8. Withstands extremely high pressures
9. Tolerant of low temperatures
10. Efficiently moves moderate volumes of water
11. Efficiently moves large volumes of water
12. Withstands prolonged storage
13. Highly resistant to charring
14. Resistant to collapse under vacuum
15. Repairable in the fire station

a hose that performs in a specific manner and possesses characteristics that maximize performance. For example, attack hose must be maneuverable and resist abrasion, large diameter supply hose must move large volumes of water efficiently over long distances, and hard suction hose must resist collapse when subjected to vacuum conditions. Forestry hose must be light and compact, and some fire extinguisher hose must be able to withstand extremely low temperatures (as when a CO_2 extinguisher is discharged). Hose must not only be constructed according to specific physical standards, but also must meet certain performance criteria. Performance requirements vary for each category of hose and are dependent on specific use. These requirements are detailed in Table 1.2.

HOW FIRE HOSE IS CONSTRUCTED

Modern fire hose is constructed by four basic methods. Each manufacturer has refined these methods to produce what they feel is a product that not only performs at the highest level, but is competitive in the marketplace against other manufacturers' products. For this reason, the methods of construction described in the following section will vary slightly from manufacturer to manufacturer.

The four basic classifications of hose, based on the method of construction, are

- Woven-jacket hose
- Rubber-covered hose
- Braided hose
- Wrapped hose

To better understand the relationship between the way hose is manufactured and the way it is used, refer to Table 1.3.

**TABLE 1.3
USE CLASSIFICATION/TYPE
OF CONSTRUCTION**

WOVEN-JACKET	RUBBER-COVERED	BRAIDED	WRAPPED
Attack	Attack	Booster	Hard Suction
MDH Relay-Supply	MDH Relay-Supply	Extinguisher	
LDH Relay-Supply	LDH Relay-Supply		
Soft Sleeve	Soft Sleeve		

Woven-Jacket Hose Construction

There are two general types of woven-jacket fire hose: unlined hose and lined hose. Each type of hose has a specific fire fighting application, especially when weight is a factor in maneuverability.

Unlined fire hose, as the name implies, has no inner tube or liner (Figure 1.13). As noted previously, it is available in two classifications: forestry hose and standpipe hose. A closely woven linen jacket serves the same function as the rubber tube in lined hose. When unlined hose is first charged, some seepage occurs until the fabric becomes saturated and swells. The hose then holds water with little leakage. In certain instances, some seepage is desirable, as with forestry hose, because it helps protect the fibers from heat damage. Although this hose is more vulnerable to abrasion damage and has a higher friction loss than lined hose, its light weight and compactness make unlined hose desirable for such purposes as wildland fire fighting.

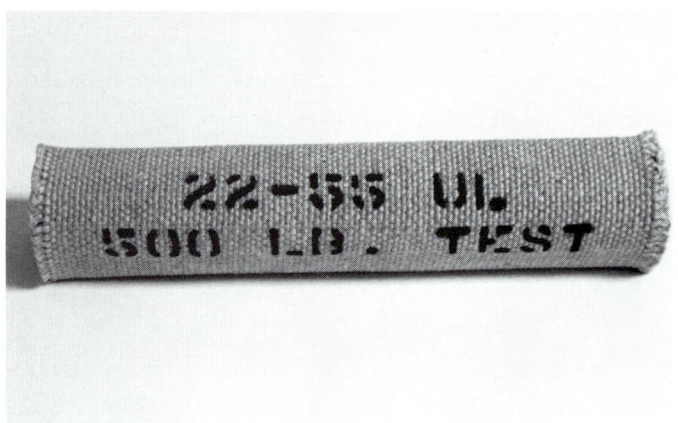

Figure 1.13 An unlined fire hose has no inner tube or liner.

Lined fire hose is by far the more widely used type of fire hose. It consists of one or more woven-fabric seamless jackets into which a rubber tube has been inserted and vulcanized (Figure 1.14). The smooth-walled tube makes the hose leakproof and reduces friction that occurs when water moves through the hose at a high velocity. Lined fire hose can withstand higher internal pressures than unlined hose and is more resistant to damage by abrasion.

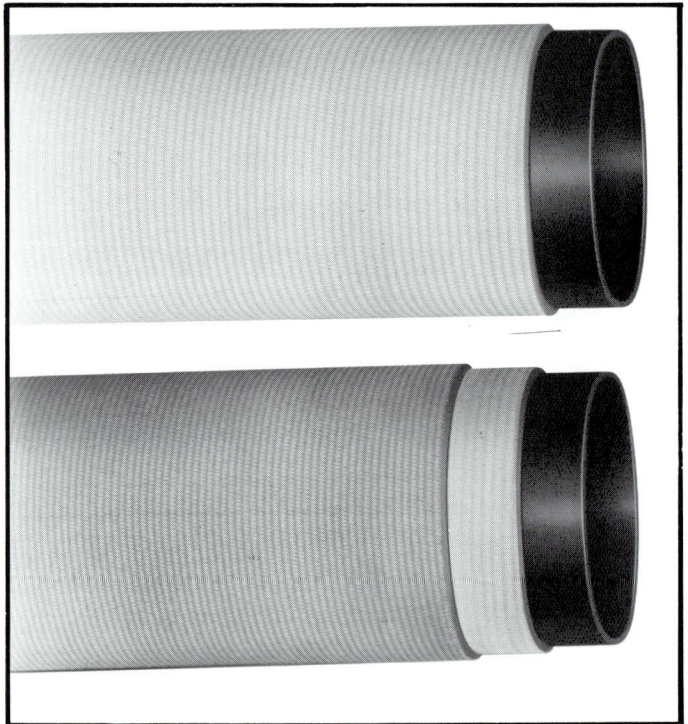

Figure 1.14 A lined fire hose has a rubber or plastic tube within one or two protective jackets. *Courtesy of National Fire Hose.*

THE INNER LINER

Essential characteristics of fire hose liners (also called tubes) are the absence of defects and pinhole leaks, good aging characteristics, and smoothness of surface. Desirable aging characteristics are required to prevent fold or edge cracking and subsequent failure in service. This feature is very important since the hose is stored during the greater part of its life. The inner liner is made of either a rubber compound or a thermoplastic.

The compounding of rubber fire hose liners involves the selection of proper ingredients to assure a satisfactory balance of all the desired properties. The first and most important of these selections is that of the basic rubber itself. Fifty years ago, natural rubber was the only choice possible. Today, however, several synthetic rubbers have become available commercially. These rubbers are actually rubber compounds made by combining pigments, processing oils, antioxidants, antiozonants, accelerators, and curative ingredients with the basic rubber to provide the desired physical properties in the finished tube. So great has been the acceptance of these rubber compounds that their nationwide use now exceeds that of natural rubber. Rubber compounds possess a much greater resistance to ozone, oil, and fuels, as well as to the overall effects of weather, than natural rubber.

Thermoplastics such as polyurethane are becoming increasingly popular in the manufacture of fire hose liners. Thermoplastic liners have five to six times more tensile strength than rubber liners, yet often weigh 40 percent less.

The liner is manufactured in either of two ways: by calendering or by extrusion. Calendering is a process in which rubber is pressed between opposing rollers to produce a flat sheet. A tube is then formed by lapping and bonding together the edges of the sized sheet to form a tube. This method has been virtually replaced by extrusion, a process in which a heated mass of rubber or plastic is forced under pressure through the orifice of an extrusion machine die to produce a continuous seamless tube (Figure 1.15). Extruded thermoplastic tubes are made in thicknesses approximately one-third that of rubber tubes. Liquid adhesives are added to the outside of the thermoplastic tubes instead of backing material.

The interior smoothness of the liner reduces the friction loss caused by passage of water through the hose at high velocity. The tube is first partially cured in its natural round shape to facilitate easy handling during the balance of the manufacturing process. After the tubes are extruded, a soft lubricating rubber backing is applied (Figure 1.16). This backing serves as an adhesive that bonds the tube to the jacket during curing. After the tube is inserted into the jacket, the final cure takes place. Heat and pressure from curing causes the backing or adhesive to bond to the inside threads of the jacket, which unitizes the tube and jacket.

Figure 1.15 An extrusion machine produces a seamless tube. *Courtesy of Snap-Tite, Inc.*

Figure 1.16 The partially cured inner liner entering this backing machine will be covered with a soft thin adhesive sheet (middle roller) and then cured inside a woven jacket by steam pressure.

THE JACKETS

The function of the jacket is to protect the inner liner and to provide strength to the hose assembly. Woven on a circular loom (Figure 1.17), the jacket has two basic elements — warp yarns and filler yarns. The warp yarns run lengthwise through the jacket; the filler yarn (sometimes called the weft yarn) runs circumferentially around the jacket, covering the warp yarns (Figure 1.18). Because the hose is subjected to a high internal pressure, and thus stress, when transporting water, the warp yarns counteract the lengthwise component of the internal stresses and the filler yarns counteract the circumferential stresses.

Cotton once ranked first in the manufacturing of fire hose jackets. Synthetic fibers, however, are now predominantly used. Several man-made fibers are used, mainly nylon and polyester. Each type of fiber has its particular advantages and disadvantages. Cotton, for instance, has good abrasion resistance, but is susceptible to mildew unless it is treated with an antimildewing fungicide. Synthetic fibers, which possess high unit strength, are more resistant to damage from chemicals and are lighter than cotton. Another advantage of synthetic is that it does not mildew even when stored wet. A disadvantage, however, is that synthetic fibers are susceptible to damage from direct contact with burning material. This is due in part to their resistance to "wicking," or absorbing water. Cotton, because it readily absorbs water, is more resistant to heat damage than synthetic. Nylon is highly resistant to abrasion and kinking, but is

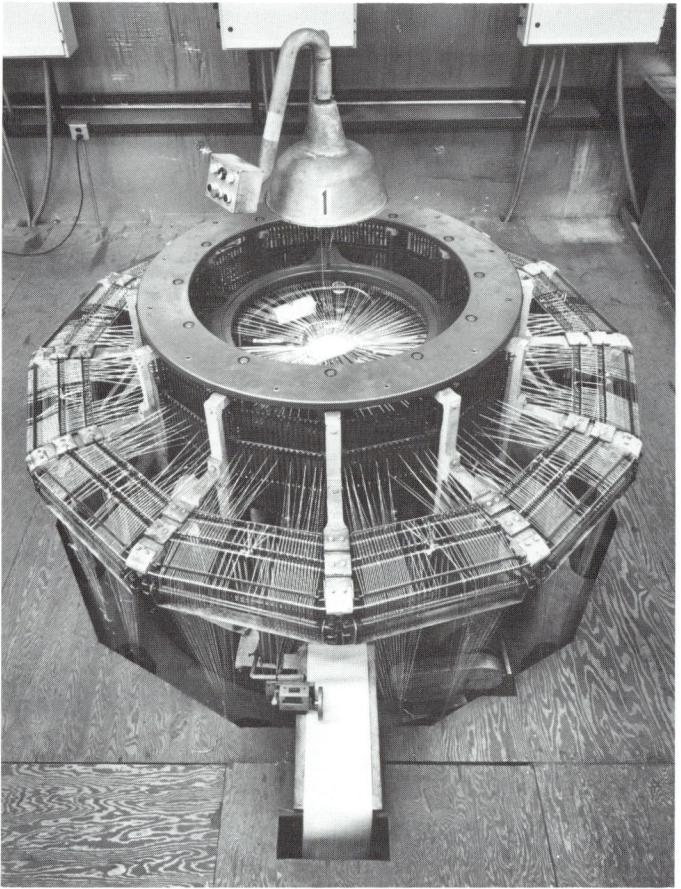

Figure 1.17 A circular loom weaves the hose jacket with warp yarns and filler yarns. *Courtesy of National Fire Hose Corp.*

also more expensive than cotton or polyester. It is used in the production of several kinds of hose, as well as in industrial hose for rack use in factories, mills, and office buildings.

Jackets may be made with the same fiber for both the warp and filler or may be made with one type of fiber in the warp and a different type of fiber

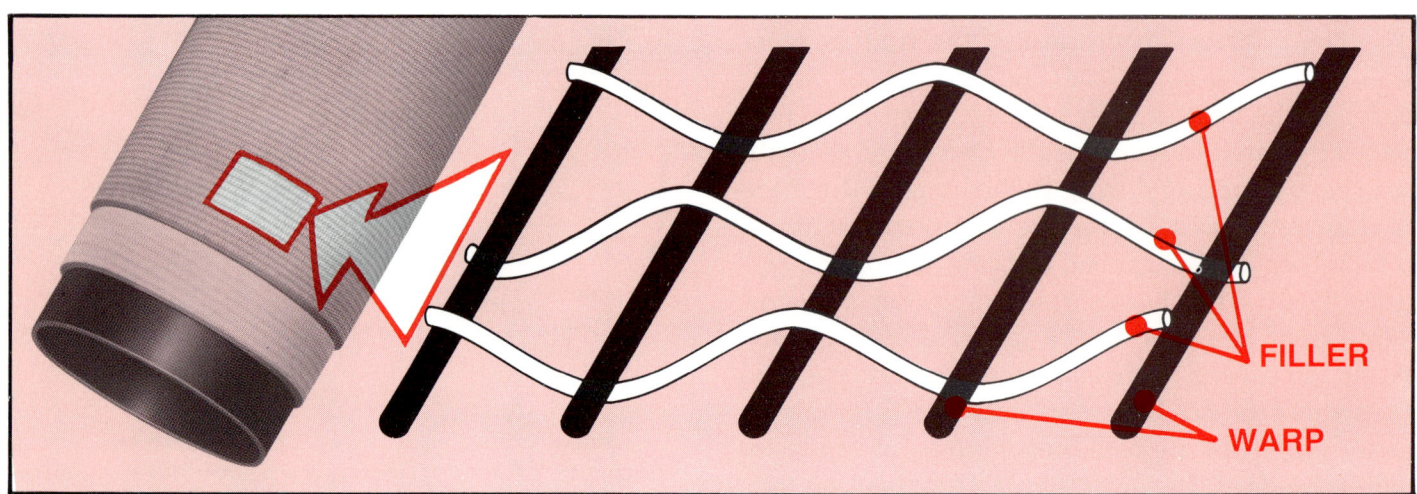

Figure 1.18 The warp yarn runs lengthwise through the jacket; the filler yarn runs circumferentially around the jacket.

in the filler. A popular combination of two different fibers is the cotton-polyester jacket. In this weave, cotton is used as the warp yarn to provide abrasion resistance and polyester is used as the filler yarn to provide burst strength. The overall result is a hose that is stronger, more flexible, and lighter than an all-cotton hose.

The most important characteristic of any fire hose is its behavior under pressure. Elongation, warping, twisting, or any distortion under pressure must be reasonably low so that desired performance can be obtained. The design and strength of the jacket is a fundamental factor in achieving this performance.

In weaving, the interlacing of the warp through the filling yarn results in a crimp in the warp yarn. The crimp, or bending, as the yarns run over and under the filling yarn detracts both from the pressure behavior and the strength of the hose. It is, therefore, desirable to keep the crimp at a minimum. The filling yarns must be woven into the jacket under uniform tension to be sure that each carries its full share of the load and keeps circumferential expansion to the minimum.

A basic pressure behavior of a woven jacket is to twist. This results from the tendency of the circular woven construction to unwind. The direction of twist in the finished hose, therefore, depends directly on the direction in which the jacket is woven. A prime requirement of all fire hose is that if it twists, it twists so as to tighten its couplings rather than to loosen them. All single-jacket and double-jacket hose is designed to meet that requirement. The two jackets in double-jacket hose are woven in opposite directions so that the tendency of the inner jacket to twist will be counteracted to a great degree by the outer jacket.

If hose is to be treated with an antimildew and water-repellent solution, the jackets are run through dipping tanks where each fiber becomes saturated with the solution. The jackets are then dried at controlled temperatures and are ready for further processing. One of the most common materials used to treat hose is chlorosulfonated polyethylene (Hypalon ®), which makes the jacket more resistant to damage from heat and abrasion.

ASSEMBLY OF THE JACKETS AND LINER

The first step in assembling the finished hose is to combine the jackets — here the inner jacket is inserted within the outer jacket. Then the semicured liner, covered with backing, is inserted into the jackets. The hose is then vulcanized in a multiple-length curing unit by injecting steam within the hose tube, which forces it out against the woven jacket. Careful attention is given to keeping the temperature and pressure at constant specified levels. During this curing process, each length of hose is placed under controlled tension so that elongation is minimized in the finished hose when under pressure.

Fire hose is "round cured" by allowing the jackets and tube to assume their natural round shape when under internal steam pressure. After the final cure, the finished hose is transported to the coupling machines for coupling prior to the hydrostatic proof test.

The final step in the manufacturing process is to roll the hose for storage and shipment. At this time the hose assumes its traditional flat shape, which is set into the hose after a period of time in the roll.

Rubber-Covered Hose Construction

In an attempt to produce a more lightweight, durable hose, some hose manufacturers have departed from the traditional production of woven-jacket hose. The general term used to describe the new type of hose now appearing on the market is "rubber-covered." As the term indicates, the hose has a rubberized cover that is mildew-proof and resists damage by abrasion and contact with chemicals.

There are a number of processes for producing rubber-covered hose. More manufacturers are entering the field each year as this type of hose gains popularity. In one process ("through-the-weave" construction), a single, circular-woven fabric tube is passed through an extrusion machine that coats the tube inside and out with a rubberized material (Figure 1.19). The woven tube is made of polyester, nylon, or a combination of these synthetic fibers. As the tube passes through the machine, a nitrile rubber compound is injected under heat and pres-

Figure 1.19 A special extrusion process produces rubber-covered hose by permeating and coating a circular-woven fabric tube, inside and out, with rubber. *Courtesy of National Fire Hose Corp.*

sure so that it permeates the tube fiber, bonding the inside and outside of the tube. The nitrile rubber thus serves as both a covering and as a smooth, leakproof liner.

Another type of rubber-covered hose is made in a three-ply process in which nitrile rubber is vulcanized to the interior surface of a woven polyester tube. This smooth inner surface acts much the same as the extruded inner tube in fabric-jacket hose: to move water with a minimum of friction loss. The rubberized tube is then coated on the outside with a protective layer of synthetic rubber (Figure 1.20).

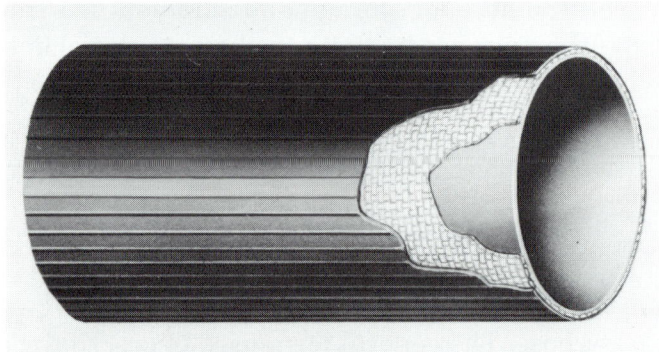

Figure 1.20 Another way to produce rubber-covered hose is to vulcanize the inside of a woven polyester tube with nitrile rubber, then coat the outside with a protective layer of synthetic rubber. *Courtesy of Snaptite, Incorporated.*

Braided Hose Construction

Braided hose is used in the manufacture of booster hose and fire extinguisher hose. In this process, a rubber liner is covered with several alternate layers of braided yarn and rubber, then vulcanized to produce a hose capable of withstanding high internal pressure (Figure 1.21). Hose constructed in this manner is often referred to as reinforced, rubber-covered, rubber-lined hose. The relatively thick covering and liner result in a rigid, noncollapsible hose that maintains an open waterway when coiled or rolled on a reel. If the hose is required to withstand extremely high pressures, the jacket is sometimes reinforced with a wire mesh.

Braided jacket construction gives hose extra strength and durability. Moisture cannot enter the jacket so no drying is required. The rubber cover also provides some protection against damage from acids or chemicals.

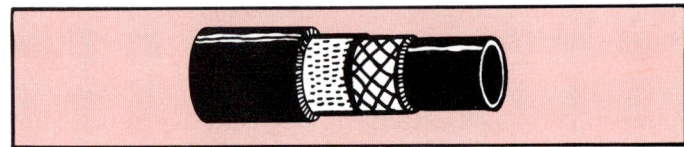

Figure 1.21 Braided hose is a rigid, noncollapsible hose made of several layers of rubber and braided material.

Wrapped Hose Construction

Hard suction (intake) hose is constructed by wrapping several layers of diagonally cut (bias-cut) fabric around an extruded rubber tube. A rubber compound is applied between the layers to hold them in place. To prevent the hose from collapsing when subjected to a vacuum condition during drafting operations, a heavy-gauge galvanized or copper wire is coiled around the hose between layers of wrapping (Figure 1.22).

Figure 1.22 A hard suction hose is reinforced with wire to prevent it from collapsing during drafting.

Fabric wrapping adds strength but makes the hard suction hose relatively heavy and only slightly flexible. The hose often requires two or three persons to couple and position it for drafting. Some hard suction hoses are made in a corrugated design, (Figure 1.23), which makes them more flexible than the older type of hard suction hose. The new design is also considerably lighter than the older style hose.

Figure 1.23 Some hard suction hoses are made in a corrugated design to make them more flexible.

Hose and Coupling Construction

HOW HOSE COUPLINGS ARE CONSTRUCTED

Coupling manufacturers make fire hose couplings out of either brass or aluminum alloy. Brass is an alloy of copper and zinc (usually composed of not less than 83 percent copper and 5 percent tin; and not more than 7 percent zinc and 3 percent lead). Couplings were once made exclusively with brass, which is both durable and resistant to corrosion, important in coastal regions where hose is exposed to salt water. It is, however, relatively heavy when compared to aluminum. Aluminum alloy (sometimes referred to as simply "aluminum") used in modern couplings is not only lighter than brass but much stronger. The strength of the alloy permits the couplings to be recoupled to hose more often than brass couplings because the hose bowl resists stretching when used with an expansion type of recoupling machine (described later in this chapter). A disadvantage of aluminum alloy, however, is that it is more vulnerable to corrosion than brass. Some manufacturers coat the alloy couplings with a hard finish to resist this corrosion.

Couplings are made by casting, extruding, or forging. A cast coupling is made by pouring molten metal into a mold. The molten metal is allowed to cool and the mold is removed, leaving a coupling that requires little further processing. Cast couplings are the weakest of the three types of couplings because the molecules of a cast metal are less densely arranged than in the same metal when extruded or forged.

Extruded couplings are stronger than cast couplings because the metal in the coupling is extruded, which requires pressure. This pressure causes the molecules of the metal to compress into a dense arrangement that gives the coupling more tensile strength than a cast coupling of the same metal (a coupling with high tensile strength is able to resist stretching forces, an important attribute when it is recoupled to hose repeatedly). With this process, the metal is extruded through a die that forms a tube with lug-shaped protrusions. The tube is then cut into sections and each section is machined to produce a finished coupling (Figure 1.24). Forged couplings are stronger than cast or extruded couplings. This process involves pounding a hot metal pellet into a forging die, which forms the metal into the desired shape. A drop-forged coupling is made by raising and dropping a drop hammer onto the metal as it rests on the forging die. Forging produces a coupling with great tensile strength because the pounding of metal compresses its molecules into an extremely dense arrangement.

Types of Couplings

Fire hose couplings are as diverse in design as the hose to which they are attached. There are, however, a few basic rules that apply to coupling design. One of the most important rules is that couplings be made to resist detachment when subjected to high internal pressures (commonly referred to as "blowing a coupling"). Mated couplings should resist becoming unconnected, not only under pressure but also when the hose is pulled, as when a charged attack line is dragged into position.

Another important design rule is that couplings have the same inside diameter as the inner surface of the hose to which they are attached. This permits a smooth flow of water through the connection without loss of pressure. There are occasions, however, when hose must be fitted with couplings that are either larger (increaser couplings) or smaller (reducer couplings) than the hose itself. This is done so that hoses of different sizes can be connected without using adapter fittings. The interior surface of a reducer coupling is tapered to

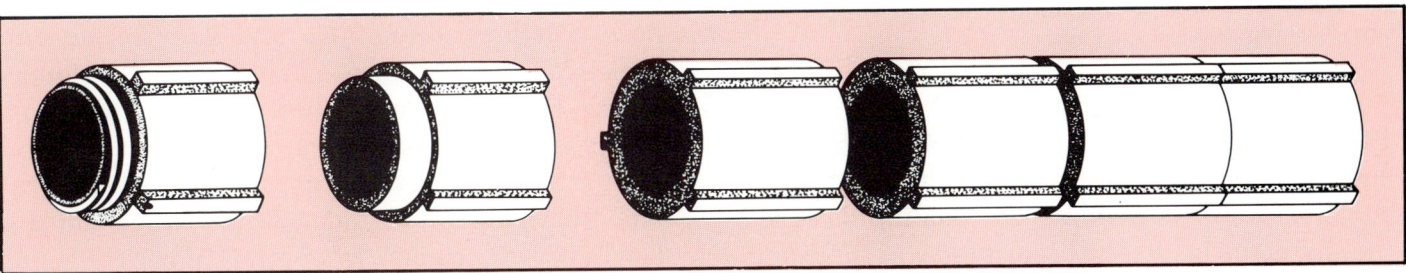

Figure 1.24 Extruded couplings are cut from an extruded tube made with lug-shaped protrusions.

minimize the pressure loss that is sure to occur when water must enter a smaller opening.

There are three basic types of hose couplings in fire service use:

- Threaded couplings
- Sexless couplings
- Snap couplings

THREADED COUPLINGS

One of the oldest coupling designs involves the casting or machining of a spiral thread into the face of two distinctly different couplings — a male and a female. A male coupling thread is cut on the exterior surface, while a female thread is on the interior surface of a free-turning ring called a swivel. The swivel permits connecting two sections of hose without twisting the entire hose. Some manufacturers make the larger sizes of couplings with either ball bearings or roller bearings under the swivel to ensure their smooth operation. Unlike common pipe threads, which are relatively fine, fire hose coupling threads are coarse, which aids in connecting the couplings quickly. A flattened angle at the end of the thread, called the blunt start, or Higbee cut, prevents cross-threading when couplings are connected (Figure 1.25).

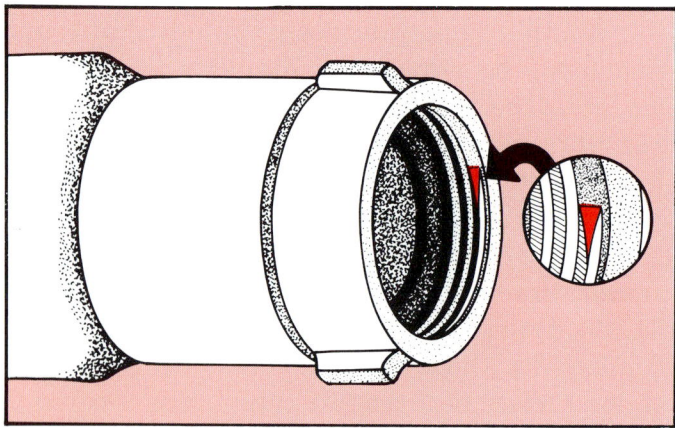

Figure 1.25 A Higbee cut at the end of the thread reduces the chance of cross-threading couplings.

A threaded coupling has several other parts. The portion that serves as a point of attachment to hose is called the shank (also called the tail piece, bowl, or shell). Depending on coupling design, if the shank fits inside the hose, it reduces the diameter of the waterway; if it fits over the outside of the hose, it does not reduce the size of the waterway. Drop-forged couplings frequently have an embossed ridge on the shank. The ridge serves to protect the swivel if a coupling is dropped or when hose is laid out from the apparatus.

Each threaded coupling is manufactured with either lugs or handles to aid in tightening and loosening connections. Lugs are located on the shank of a male coupling and on the swivel of a female coupling. They aid in grasping the coupling when making and breaking coupling connections. Connections may be made by hand or with spanners, which are special tools that fit against the lugs. There are three types of lugs: pin, recessed, and rocker. Pin lugs, usually found on couplings of older hose, resemble small pegs (Figure 1.26). Although still available, pin-lug couplings are not

Figure 1.26 Pin lugs resemble small pegs.

commonly ordered with new hose because of their tendency to hang up when hose is dragged over objects. Booster hose normally has couplings with recessed lugs, which are simply shallow holes drilled into the coupling (Figure 1.27). This lug design prevents abrasion that would occur if the hose had protruding lugs and was wound onto reels. Modern threaded couplings have rounded rocker lugs. Unlike pin lugs, the rounded shape of rocker lugs helps prevent snagging. On the couplings, one of the rocker lugs on the swivel is scalloped with a shallow indentation, called the Higbee indicator, to mark where the Higbee cut begins (Figure 1.28). This indicator aids in matching the

Hose and Coupling Construction **19**

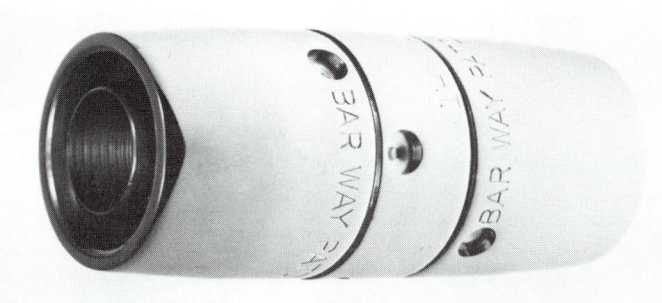

Figure 1.27 Recessed lugs are shallow holes bored into the couplings. *Courtesy of Bar-Way Manufacturing Company.*

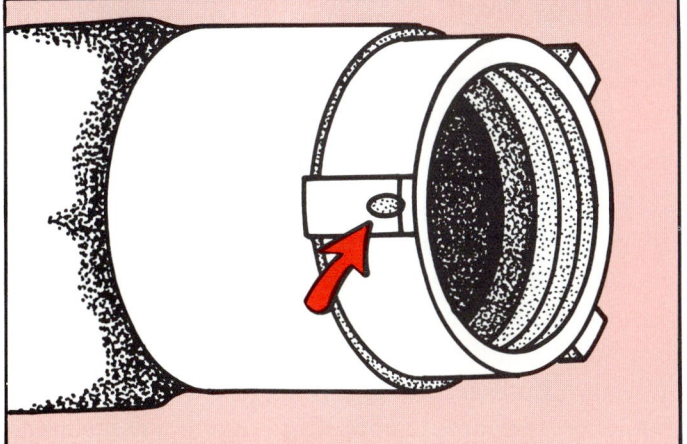

Figure 1.28 The Higbee indicator marks where the Higbee cut begins on the couplings.

male coupling thread to the female coupling thread, which is not readily visible.

Handles are used primarily on intake hose and are located only on the swivels (Figure 1.29). They aid in tightening the large coupling by hand when connecting the hose to a pump intake valve. The coupling can be further tightened by striking the handles with a rubber mallet.

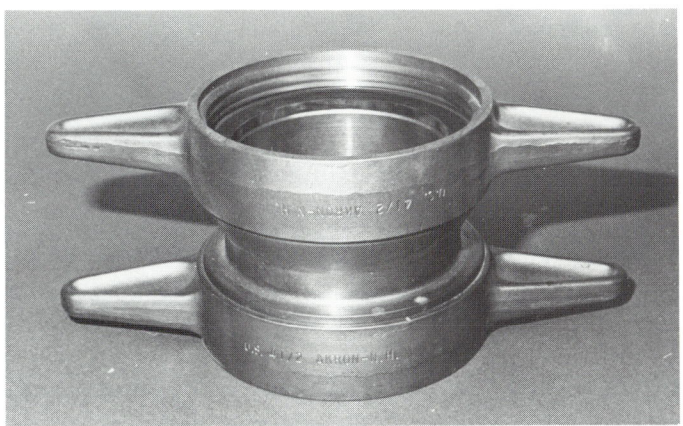

Figure 1.29 Handles are sometimes built into large couplings so they can be tightened by hand.

Each section of hose with threaded couplings has a male coupling at one end and a female coupling at the opposite end. Together, the two couplings are referred to as a set. The set is also referred to as a three-piece coupling (the male coupling is considered one piece, and the female coupling is considered two pieces because it also has a swivel). Table 1.4 provides information about the sizes of threaded couplings available for common hose sizes.

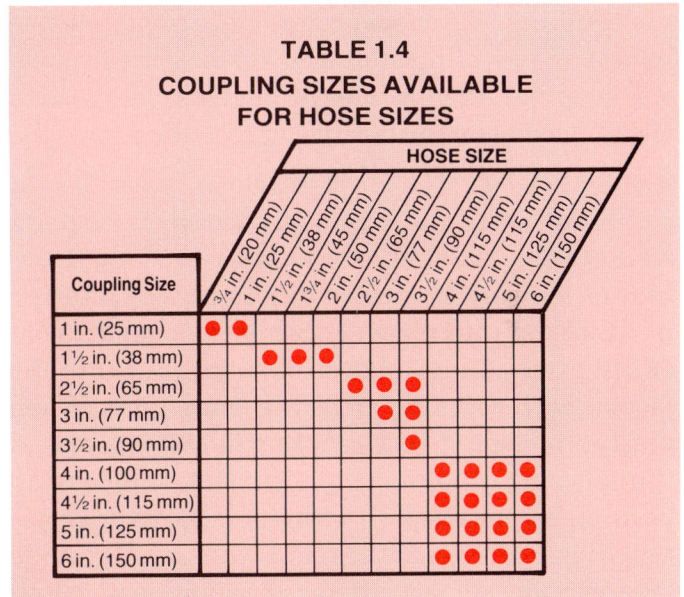

TABLE 1.4
COUPLING SIZES AVAILABLE
FOR HOSE SIZES

Coupling Size	¾ in. (20 mm)	1 in. (25 mm)	1½ in. (38 mm)	1¾ in. (45 mm)	2 in. (50 mm)	2½ in. (65 mm)	3 in. (77 mm)	3½ in. (90 mm)	4 in. (100 mm)	4½ in. (115 mm)	5 in. (125 mm)	6 in. (150 mm)
1 in. (25 mm)	●	●										
1½ in. (38 mm)			●	●	●							
2½ in. (65 mm)						●	●	●				
3 in. (77 mm)							●	●				
3½ in. (90 mm)								●				
4 in. (100 mm)									●	●	●	
4½ in. (115 mm)									●	●	●	
5 in. (125 mm)									●	●	●	
6 in. (150 mm)										●	●	●

SEXLESS COUPLINGS

While threaded couplings have been in fire service use for over a century, another type of coupling has appeared in North America during the past two decades — the sexless coupling. This term means that there are no distinct male and female components, so both couplings are identical. There are several advantages, as well as disadvantages, to using hose with this type of coupling:

- Hose can be quickly connected, but spanners should be used to ensure a complete connection. This slows the connecting operation.

- Hose can become uncoupled, often suddenly and violently, if a complete connection has not been made.

- The possibility of cross-threading is eliminated.

- There is no need for double male or double female adapters, thus hose can be laid in either direction.
- An adapter is required at the hydrant, which lengthens the hook-up time.

There are two kinds of sexless couplings: quarter-turn and Storz. The quarter-turn coupling has two hooklike lugs on each coupling (Figure 1.30). The lugs, which are grooved on the underside, extend beyond a raised lip or ring on the open end of the coupling. When the couplings are mated, the lugs of one coupling slips over the ring of the opposite coupling, then rotates 90 degrees clockwise to lock. A gasket on the face of each coupling seals the connection to prevent leakage.

Storz couplings, commonly found on large diameter hose, are similar to quarter-turn couplings in that they are connected by joining and rotating until locked into place. Locking components consist of grooved lugs and inset rings built into the face of each coupling swivel (Figure 1.31). When mated, the lugs of each coupling fit into recesses in the opposing coupling ring, then slide into locking position behind the ring with a one-third turn rotation. External lugs at the rear of the swivel provide leverage for connecting and disconnecting couplings. They also indicate complete connection by coming into alignment when the couplings lock into place.

Figure 1.31 Storz couplings have interlocking grooved lugs.

SNAP COUPLINGS

Another coupling design available to the fire service, although not in widespread use, is the snap coupling. Snap coupling sets have both a male and female component. The female coupling has a shallow bowl that fits over the end of the male coupling. When a connection is made, two spring-loaded hooks on the female coupling engage a raised ring around the shank of the male coupling (Figure 1.32).

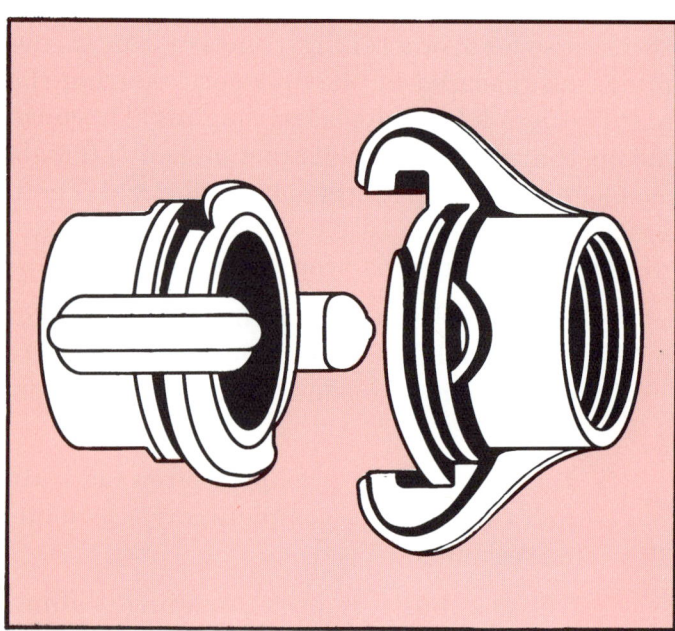
Figure 1.30 Quarter-turn couplings have hooklike lugs.

Figure 1.32 Snap couplings interlock when the two spring-loaded hooks on the female coupling engage a ring on the shank of the male coupling.

As with sexless couplings, the primary advantage with snap couplings is that hose can be connected more quickly than with threaded couplings. A disadvantage, however, is that the coupling hooks tend to hang up on curbs and other obstacles when the hose is pulled across the ground.

HOW COUPLINGS ARE ATTACHED TO HOSE

As stated earlier, a primary requirement of coupling design is that the coupling be firmly attached so that it resists detachment when the fire hose is pressurized and when the hose is pulled or dragged. There are five different components, and thus methods, for attaching couplings to fire hose:

- Expansion rings
- Screw-in expanders
- Collars
- Tension rings
- Banding

Expansion Ring Method

One of the oldest ways to attach a threaded coupling to a fire hose involves the use of a malleable metal band called an expansion ring. The expansion ring, which is slightly smaller than the hose, is placed inside and flush with the end of the hose. The hose is pushed into the coupling hose bowl, then the ring is expanded against the hose with a hydraulic expanding device. This compresses the hose tightly against the inner surface of the coupling. Figure 1.33 illustrates a double-jacket hose with an expansion ring coupling. The ex-

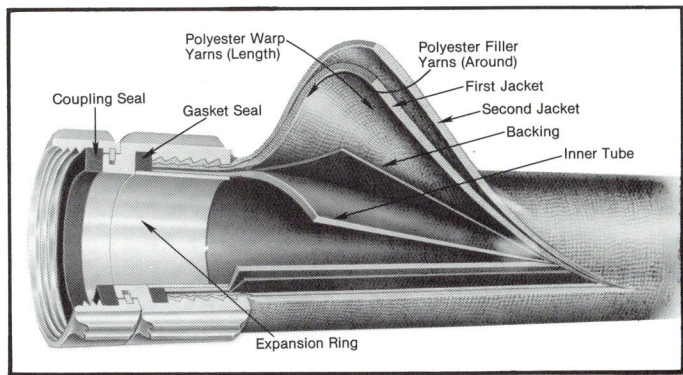

Figure 1.33 A double-jacket fire hose with an expansion ring coupling. *Courtesy of Badger-Powhatan.*

panded ring is the same diameter as the hose liner so that it does not obstruct the waterway. This method is not only used at the factory where hose is made but also in the fire station shop to recouple hose.

Screw-In Expander Method

Some types of hose, particularly rubber-jacket booster hose, have threaded couplings attached with expanders that are screwed into place. Unlike the metal expansion ring described above, the screw-in expander is actually an integral component of the coupling, which is made of two pieces, the shell and the expander (Figure 1.34). The inside surface of the shell is serrated to prevent slippage of the attached hose. One end of the expander, rather than the shell, contains the coupling threads. The opposite end of the expander is also threaded so that it may be screwed into the coupling shell. The coupling is attached to the hose by placing the shell over the hose end, then screwing

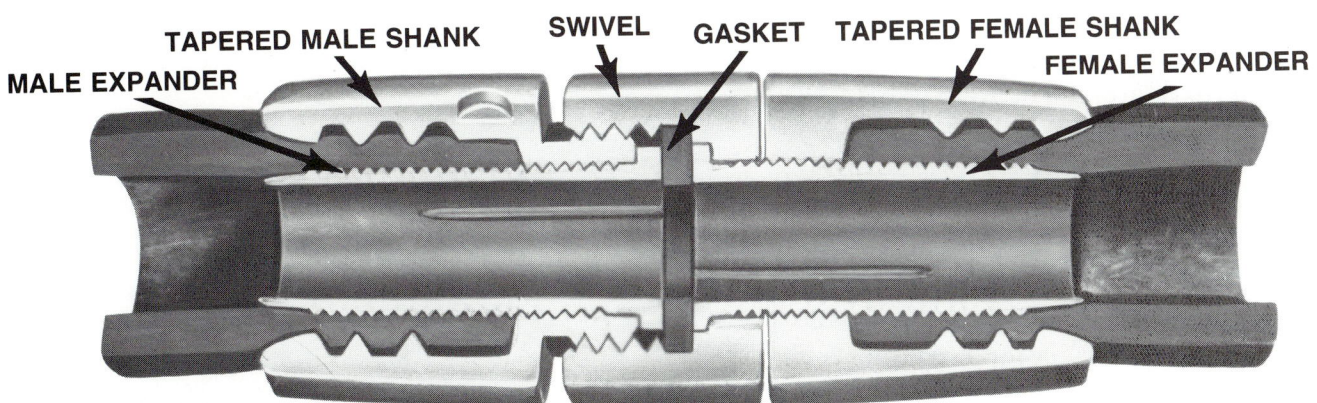

Figure 1.34 The screw-in expander coupling is composed of a shell and an expander. *Courtesy of Bar-Way Manufacturing Company.*

the expander into the hose until it seats against the face of the shell. This compresses the hose tightly against the serrations inside the shell.

Collar Method

Another coupling designed with a shank that fits inside the hose is fastened to the hose with a collar that bolts into place. This is one of the most simple ways to attach a coupling because no equipment except a wrench is required. This coupling is most often used with large diameter hose (Figure 1.35). With this method, the hose is slipped over

Figure 1.36 The tension ring method compresses the hose material against inside grooves on the coupling shank. *Courtesy of Snaptite, Inc.*

Figure 1.35 The collar method is often used to attach couplings to large diameter hose. *Courtesy of National Fire Hose.*

the coupling shank, which is serrated, and then a two-piece or three-piece collar is fastened in such a way that it compresses the hose against the shank. It should be noted, however, that with this type of attachment the waterway through the coupling is reduced (approximately ½ inch [13 mm] in a 4-inch [100 mm] hose).

Tension Ring Method

A method similar to the collar method for attaching couplings utilizes a tension ring (Figure 1.36). In this case, the coupling shank is made with two grooves around its outer circumference. As with the collared coupling, this shank fits inside the hose end. A nylon sleeve with inside ridges that correspond to the grooves on the shank is placed on the hose directly over the shank grooves. The tension ring is then placed over the nylon sleeve and tightened with allen-head bolts. As the bolts are tightened, the ridges on the inside of the nylon sleeve compress the hose material against the grooves on the coupling shank, making a tight-fitting attachment.

Banding Method

A coupling rarely found in the fire service, but nevertheless still in use, is attached to fire hose with tightly wound bands of narrow-gauge wire or with steel bands (Figure 1.37). In this case, the coupling shank is designed to fit inside the end of the hose rather than over the outside, as most other types of couplings. The coupling shank is made with ridges and grooves so that the wire or steel band seats into the grooves. The ridges prevent the coupling from detaching under pressure.

Figure 1.37 One type of coupling is attached by tightly banding narrow gauge wire or steel bands over the hose and coupling shank.

Hose and Coupling Construction 23

Chapter 1 Review
Answers on page 233

MULTIPLE CHOICE: Circle the correct answer.

1. Most fire service hose is _____.
   A. braided
   B. wrapped
   C. woven-jacket
   D. rubber-covered

2. Fire extinguisher hose is made in essentially the same way as _____.
   A. booster hose
   B. forestry hose
   C. intake hose
   D. standpipe hose

3. Which type of fiber is most widely used for woven-jacket fire hose?
   A. Cotton
   B. Polyester
   C. Nylon
   D. Synthetic

TRUE-FALSE: Mark each statement true or false. If false, explain why.

4. Most fabric-jacket fire hose is single-jacketed.
   ☐ T ☐ F _____

5. When using a domestic standpipe system as a water source to combat a fire, firefighters should connect fire department hose rather than use the hose found preconnected to the system.
   ☐ T ☐ F _____

6. Booster hose is a rubber-covered hose made of several layers of braided, rubberized material.
   ☐ T ☐ F _____

7. Booster hose is usually folded into a portable pack or bundle, which is carried on the apparatus.
   ☐ T ☐ F _____

8. Booster hose is manufactured in 2½- and 3-inch (65 mm and 77 mm) diameter sizes.
   ☐ T ☐ F _____

24 HOSE

9. Soft sleeve hose is usually used to transfer water from a hydrant to the pump intake, while hard suction hose is usually used for drafting.
   ☐ T ☐ F _____
   _____

10. Curing is the application of heat and pressure to a fire hose liner to set its shape, increase its smoothness, and bond it to the outer jacket.
    ☐ T ☐ F _____
    _____

11. The same type of fabric must be used for both the warp yarn and filler yarn in a woven-jacket fire hose.
    ☐ T ☐ F _____
    _____

12. A primary requirement of a fire hose coupling is that it resists detachment when the hose is pressurized or dragged.
    ☐ T ☐ F _____
    _____

MATCHING: Write the correct letter in the space provided.

13. Match terms associated with hose types to their descriptions.

    _____ Attack
    _____ Intake
    _____ Extinguisher
    _____ Supply

    A. A hose used in a handline to control and extinguish fire.
    B. A hose designed to move large volumes of water at low pressure.
    C. A hose used to connect a pumper to a nearby water source.
    D. A hose designed to transport extinguishing agents from the container to the nozzle.

14. Match terms associated with hose construction to their descriptions. Write the correct letter in the space provided.

    _____ Wrapped
    _____ Braided
    _____ Rubber Covered
    _____ Unlined woven-jacket
    _____ Lined woven-jacket

    A. A hose constructed of one or more fabric jackets into which a rubber tube has been inserted.
    B. A hose constructed of a woven tube covered on both the inside and outside with rubber.
    C. A hose constructed by folding several layers of fabric around a rubber tube with heavy-gauge wire coiled between the fabric layers.
    D. A hose constructed only of a fabric tube.
    E. A hose constructed by covering a rubber liner with several alternate layers of braided yarn and rubber.

FILL IN THE BLANK: Fill in the blanks with the correct values.

15. Conventional extinguisher hose is used for extinguishers that discharge at pressures no greater than _____ psi (_____ kPa), while high-pressure extinguisher hose can be used with pressures up to _____ psi (_____ kPa).

SHORT ANSWER: Answer each item briefly.

16. What is the major use for forestry hose?

17. Is forestry hose single-jacketed, multiple-jacketed, or can it be either?

18. Is forestry hose lined, unlined, or can it be either?

19. Some forestry hose is designed to leak water through to the outer jacket. What advantage is gained by this feature?

20. Is standpipe hose usually single-jacketed or multiple-jacketed?

21. Is standpipe hose lined, unlined, or can it be either?

22. What is the most common size for standpipe hose?

23. What is the major disadvantage of standpipe hose?

24. Which is generally larger in diameter, supply hose or attack hose?

26 HOSE

25. Pressure loss seriously depletes the water flow when medium diameter hose is used for long distance water supply. What causes this pressure loss?

   _____
   _____

26. Why is it preferable to use hard suction hose rather than soft sleeve hose when drafting from an open water source?

   _____
   _____

27. Comparing fire hose liners made of thermoplastic and rubber, which has more tensile strength?

   _____
   _____

28. Comparing fire hose liners made of thermoplastic and rubber, which is lighter in weight?

   _____
   _____

29. What are calendering and extrusion?

   _____
   _____

30. What is the purpose of the lugs on threaded couplings?

   _____
   _____

LISTING

31. List two methods of compensating for the inadequacies of medium diameter hose when it must be used to supply large volumes of water.

   A. _____
   B. _____

32. List the five sizes of large diameter supply hose.

   A. _____
   B. _____
   C. _____
   D. _____
   E. _____

Hose and Coupling Construction

33. List the three methods of manufacturing couplings, with the strongest type listed first and the weakest last.

    A. _____

    B. _____

    C. _____

34. List the three basic designs of hose couplings in fire service use.

    A. _____

    B. _____

    C. _____

35. List the two kinds of sexless couplings.

    A. _____

    B. _____

36. List the five methods for attaching couplings to fire hose.

    A. _____

    B. _____

    C. _____

    D. _____

    E. _____

DISCUSSION QUESTIONS

If cost were not a factor, what type of fire hose would you equip your apparatus with? Why?

_____

_____

_____

_____

If cost were a factor, what type of fire hose would you equip your apparatus with? Why?

_____

_____

_____

_____

LEARNING ACTIVITY

Identify the amount and type of hose on the apparatus at your fire station.

2

Care, Maintenance, and Testing

30 HOSE

This chapter provides information that addresses performance objectives described in NFPA 1001, *Fire Fighter Professional Qualifications* (1987), particularly those referenced in the following sections:

Fire Fighter I

3-13 Fire Hose, Nozzles, and Appliances

3-13.5

Fire Fighter II

4-13 Fire Hose, Nozzles, and Appliances

4-13.3

Fire Fighter III

5-13 Fire Hose, Nozzles, and Appliances

5-13.1

Chapter 2
Care, Maintenance, and Testing

Fire hose, like other fire fighting equipment, must be used properly during fire fighting operations, as well as during training and other activities. When not in use, it must be well maintained so that it can be relied on to function without failure. This requires that it be cleaned, inspected, and tested on a regular basis. An invaluable aid to carrying out this function has been provided through a standard published by the National Fire Protection Association (NFPA). Much of the information presented in this chapter is based on that standard: NFPA 1962 *Standard for the Care, Use and Maintenance of Fire Hose Including Connections and Nozzles.*

CAUSES AND PREVENTION OF HOSE DAMAGE

During initial fire attack, little effort is usually given to protecting hose from injury. After the fire is contained, however, there is little reason not to take preventive measures to protect this essential equipment from damage. During overhaul operations and when "picking up" after fires, particular care should be taken when moving hose through debris from the fire. This is a time when hose can be quickly damaged by contact with embers, broken glass, and other objects.

Most fire hose is made of a woven material and a rubberized coating or rubber liner. These components are vulnerable to damage in a number of ways: by the mechanical action of objects, by exposure to heat, by the action of mold and mildew, and by chemical contact.

Mechanical Damage

Mechanical damage occurs when an object contacts the hose somewhere along its length and cuts, abrades, tears, or stresses the jacket and underlying material. This type of damage and ways to prevent or lessen it are discussed in the following paragraphs.

Hose is damaged when it is pulled over sharp edges such as cornices, parapets, and windowsills. Avoid this damage by using a hose roller (Figure 2.1) or improvised padding, such as a bundled salvage cover. While these procedures may not always be practical during the initial stages of fire fighting, they should become routine during such activities as overhaul operations and training evolutions.

Figure 2.1 A hose roller prevents abrasive damage to hose.

Hose may be damaged when it is advanced through window openings containing broken glass fragments. It only takes a few seconds to completely break out glass fragments from window openings. Use a spanner or axe to sweep the window sash clear of glass shards (Figure 2.2). This not only prevents glass from cutting the hose jacket, but ensures the safety of personnel climbing through the opening.

Figure 2.2 Clear a window frame of glass shards to prevent damage to the hose jacket.

Mechanical damage occurs when hose is dragged through debris. Debris usually contains sharp-edged objects, such as glass, nails, and metal, each of which can cause cuts or abrasions to the hose jacket. Route hoselines around debris, move debris from areas where hose must be moved, or cover the debris with canvas or other protective material to prevent damage to hoselines.

Tools or equipment carried on top of the hose bed may damage hose. Store tools and equipment, even temporarily, in places other than on top of hose (Figure 2.3). While the tools may cause little damage from simply laying on a hose bed, they can do great damage if hose is pulled from the hose bed.

Hose may be abraded when it is dragged over rough surfaces. This type of damage actually occurs most often during nonemergency activities such as training and hose maintenance. When hose is dragged over rough surfaces, such as asphalt or concrete pavement, the outer jacket receives a mild abrasion. If this happens repeatedly, the outer jacket is significantly weakened. To avoid this damage, carry the hose in rolls or loose folds rather than dragging it. If hose must be dragged, avoid surfaces that are exceptionally abrasive, such as graveled roadway and rocky ground, and drag it in with the flat side down, rather than on edge. Hose edges are particularly susceptible to abrasive damage when dragged.

Hose suffers damage when it is dragged on the roadway behind moving apparatus. Both the hose jacket and couplings suffer from this kind of abuse. It can occur when hose hangs up in the bed during a hose laying operation or when a section of hose vibrates loose from the load when the apparatus is in motion. Prevent hangups from occurring by not loading the hose so tightly that couplings lodge against adjoining couplings or against hose bed rails. Prevent hose from accidentally dislodging during travel by firmly securing the end coupling when loading the hose bed.

Figure 2.3 The hose bed should be free of tools and other equipment.

Care, Maintenance, and Testing   **33**

Damage occurs when hose is run over by vehicles. When a vehicle runs over a section of fire hose, damage may not be immediately recognizable. In this case, the inner liner and inner jacket, which are bonded together, may become unbound because of stresses caused by the weight of the vehicle. The liner may also become cracked at the sharp bends on the edges. This damage is usually not apparent until the hose is pressure tested. Prevent this damage by laying hose to one side of the street so that vehicles are not forced to drive over it. Provide traffic control at the fire scene to prohibit non-emergency vehicles from entering the area where hose is laid. If it becomes necessary for emergency vehicles to drive over hose, drive over charged hose rather than flat hose to minimize damage. A tarp laid over large diameter hose will prevent the hose from skidding ahead of the tires if the vehicle crosses at an angle (Figure 2.4). When hose is placed where vehicles must repeatedly drive over it, lay hose bridges or improvised ramps at crossing places to bear the weight of the vehicles.

Hose may be chafed by vibration from the pumper. Some apparatus vibrates so much that the hose connected to it, particularly the intake hose, chafes where it touches the street surface. Use a chafing block at the point where the hose contacts the ground (Figure 2.5). When hard suction hose is carried preconnected, provide padding at contact points between the hose and supporting brackets to prevent chafing.

Hose may be damaged when dirty hose is reloaded on the apparatus. Dirt and grit on the hose abrades the jacket fibers, much like sandpaper. This action is increased by apparatus vibration when moving. To prevent this problem, clean and

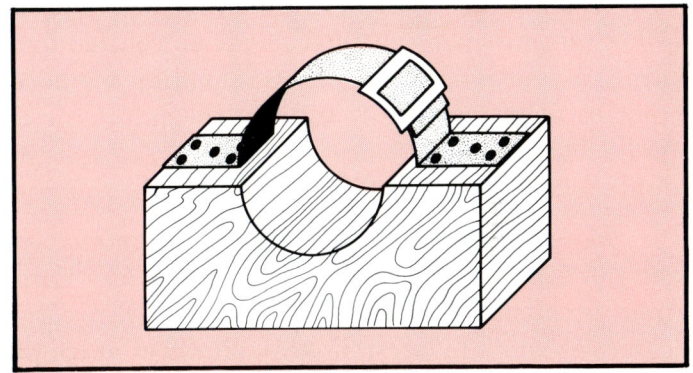

Figure 2.5 A chafing block prevents wear to intake hose caused by pump and engine vibration.

Figure 2.4 Drive over a tarp-covered LDH at an angle. The tarp will prevent the hose from skidding in front of the tires.

dry dirty hose immediately upon return from an incident. Reload the hose bed with clean hose only.

Hose may become worn if it is loaded on edge. When hose is loaded on edge, such as in a horseshoe or accordion load, it wears on the edges. Because 100 percent polyester hose and LDH are particularly susceptible to such wear, NFPA 1962 and fire hose manufacturers recommend that this hose be loaded flat.

Damage occurs if sharp folds are left in the hose for long periods. Tightly loaded hose that remains in a hose bed for long periods can develop cracks where there are sharp folds in the hose. NFPA 1962 recommends that hose loads be rotated four times a year if not used. When reloading the hose, refrain from packing it tightly; relocate folds at previously unfolded places in the hose. Stagger the folds so that they are not tightly packed.

Hose may be damaged if a water hammer occurs. A sudden increase in pressure, or water hammer, is caused by closing nozzles too quickly. This can cause hose to burst at weak points or at couplings. Prevent pressure damage by closing nozzles and valves slowly. Pressure control devices such as governors and relief valves can also prevent surges in hoseline pressures. Use an in-line relief valve to protect large diameter hose against water hammer.

Damage can occur if hose is left hanging in a hose tower for extended periods. The inner lining of hose can become damaged at the point that it hangs over the support peg if left for long periods of time. Prevent this damage by removing the hose as soon as it dries. If the hose must remain in the tower for prolonged periods, change the hose/peg contact point periodically.

Heat and Cold Damage

Heat damage to hose occurs when it contacts fire and hot objects such as found in fire debris. This causes charring, melting, or weakening of the covering or of the jacket fibers. It will also dry the lining, which promotes cracking. Similar damage can occur when hose is near apparatus engine exhaust pipes. Prevent this damage by simply routing hose away from sources of extreme heat whenever possible. Hose can also become damaged when exposed to extreme ambient temperatures for prolonged periods such as when left in a hose dryer or in direct sunlight on an extremely hot day. This causes deterioration of the liner. Prevent this damage by storing hose inside as soon as it is completely dry.

Cold damage occurs when water within the hose, as well as on the outside hose surfaces, becomes frozen. Prevent internal freezing in hose during intermittent use by allowing some water to flow through the nozzle at all times (colder temperatures require greater flows). Maintain water flow in intake hose by circulating water from the hydrant through the pump, discharging it through a drain-off hose that routes water down a gutter or to a place away from the apparatus. Immediately drain and roll hose that is no longer needed for fire fighting. Prevent couplings from leaking and thus freezing by tightening all connections. In extremely cold conditions, prelubricate couplings with antifreeze to help prevent freezing.

When hose becomes frozen in ice, there are three ways to remove it: melt the ice with a steam-generating device to remove the hose; chop the hose loose with axes; or leave the hose until the weather warms enough to melt the ice. When chopping hose out of ice, make all cuts in the ice well away from the hose to reduce the chance of the axe blade glancing into the hose fabric. Break frozen connections by pouring hot water over towel-wrapped couplings, or place couplings near apparatus exhaust pipes (avoid inhaling exhaust fumes). Using a propane torch to thaw couplings is *not* recommended because overheating can damage the hose and the gaskets. Do not fold frozen hose, as this will almost certainly cause damage. Before placing thawed hose back in service, pressure test it to ensure no damage has occurred.

Mildew and Mold Damage

When hose jackets are woven from organic fibers, such as cotton or flax, they are susceptible to attack from fungus, which is a parasite that feeds on dead organic matter. The fungus is commonly referred to as mildew or mold. The ideal condition for hose to mildew is when there is moisture in the hose jacket fibers, whether the hose is in storage racks or in the hose bed, and evaporation is inhi-

Care, Maintenance, and Testing **35**

bited by a lack of air flow. Mildew weakens the hose jacket as the fungus consumes the fibers.

This type of damage can be avoided by doing the following:

- Make sure hose is *completely* dried before storing or loading.
- Cover hose beds with water repellent covers to keep loads dry during inclement weather (Figure 2.6).
- Check hose in racks and hose beds periodically. Even if not visible, a musty smell is often an indicator that mildew is hidden somewhere within the hose.
- Ventilate all areas where hose is kept, including apparatus hose beds.

Figure 2.6 Water repellent covers keep hose loads dry.

If mildew is discovered on hose, immediately wash the hose. Scrub the jacket with a very mild soap solution and dry completely. Inspect the hose section within the next few days after treatment for the reappearance of mildew.

Chemical Damage

Many chemicals, in both liquid and gaseous form, are injurious to fire hose. Motor oil, found in some quantity on most streets and highways where hose is laid, is an example of a petrochemical that will penetrate the woven jacket and produce a solvent action on the rubber lining. This action is even more drastic with gasoline. Battery acid is another chemical that causes significant damage to hose by destroying the jacket fibers.

Some of the following recommended practices will help prevent chemical damage:

- Avoid laying hose near curbs, where oil, gasoline, and battery acid may accumulate from parked automobiles. Hose laid in gutters may also come in contact with fire fighting runoff water, which could contain harmful chemicals.
- Thoroughly scrub hose suspected of having contacted acid with a solution of bicarbonate of soda and water (Figure 2.7). The sol-

Figure 2.7 Scrubbing hose with bicarbonate of soda and water will neutralize acid on the outer jacket.

ution will neutralize the acid. Bubbling on the jacket surface, which occurs when acid and bicarbonate react, will indicate the presence of acid in the fibers.

- Test and inspect all inactive hose periodically.
- Test hose if there is any suspicion of chemical damage.

CAUSES AND PREVENTION OF COUPLING DAMAGE

Couplings become damaged usually through rough handling such as being dropped or run over by vehicles. Dropping a coupling, especially one made of brass, may cause the coupling to become misshapen, or out-of-round. This type of damage is more likely to occur when the coupling is disconnected. Swivels and male threads are particularly susceptible to damage by being dropped. Prevent this damage by simply taking care when handling hose.

LAY HOSE OUT OF THE PATH OF VEHICLES WHEN POSSIBLE TO MINIMIZE DAMAGE TO HOSE AND COUPLINGS

Few couplings can withstand the weight of a vehicle if run over. When a tire rolls over a coupling connection, the usual result is that the couplings are slightly flattened, becoming out-of-round. This is easily detected because the swivel on the female coupling will no longer spin freely, and the male coupling will not accept a female coupling. If vehicle tires pass over the hose immediately adjacent to a coupling, the hose might be pulled partially from the coupling shank. When this happens, the coupling will appear to be cocked at an angle on the hose. Prevent this damage by laying hose out of the path of vehicular traffic or by pro-hibiting vehicle access where hose has been laid. If vehicles must run over hose and hose bridges are not available, guide the vehicles over the hose well away from any couplings.

When a section of hose is dragged with couplings trailing behind, there is always a chance of damaging a coupling, especially on rough surfaces. Male threads can wear down or become chipped, and swivels on female couplings can become misshapen. If it becomes necessary to drag hose, fold it in half and carry both couplings.

Screw threads on couplings can be damaged if cross-threaded or if mismatched with couplings of another thread type. Use the Higbee indicator to help avoid cross-threading. If it is difficult to mate threads, turn the swivel counterclockwise against the male thread until a distinct click is heard, then turn the swivel clockwise to begin threading the two couplings together. If it becomes difficult to turn the swivel when connecting couplings, suspect that either the couplings are out-of-round, that threads are damaged, or that the threads do not match. Never force threads, however. Take action to correct the problem.

Coupling Repair

When a coupling is severely misshapen, as when run over by a vehicle, it must be removed and replaced. If the damage is less severe, however, the coupling can sometimes be restored to near its original shape. This will depend on a number of factors that include the degree and place of distortion, the type of metal in the coupling, and the coupling design. A brass screw-thread coupling, for example, can be gently hammered back into shape if it is the type that attaches with an expansion ring. The coupling can be placed on the mandrel of an expansion machine and tapped while under slight expansion pressure. If done properly, the coupling can be restored to near original shape.

When threads have been damaged, they can sometimes be repaired with a file or with a thread tap or die tool. Check for irregularities on the thread surface and, if needed, remove burrs and abrasions with a fine three-cornered file (Figure 2.8). If the thread is more severely damaged, use a tap or die of the proper size to restore the thread surface.

Care, Maintenance, and Testing 37

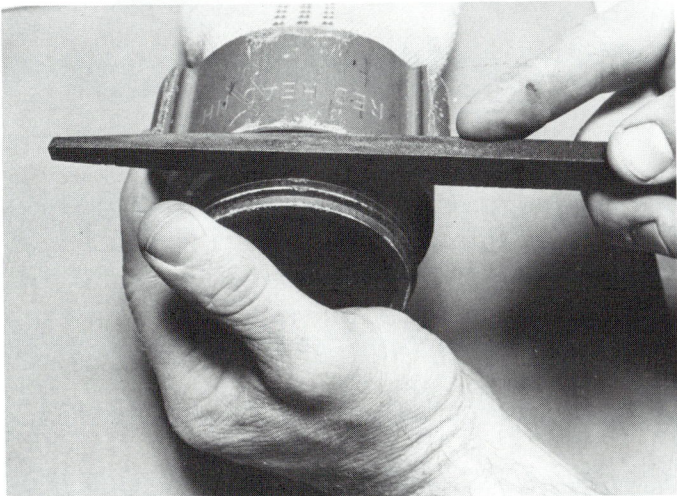

Figure 2.8 Remove burrs and abrasions on the thread surface with a three-cornered file.

GENERAL CARE AND MAINTENANCE OF HOSE

Hose is one of the most essential types of equipment carried on a fire fighting apparatus. To be dependable in every emergency, it must be maintained on a regular basis. Fire hose rarely stays clean during fire fighting. It often takes great abuse, especially in intense fire situations where it becomes difficult to follow ideal hose care rules. Cleaning the hose, therefore, is very important to prolonging the life of each section of hose.

Washing, Drying, and Storage

Hose washing is a laborious, time-consuming job that can be simplified by using such devices as a commercial hose washing machine or a jet-spray washer. A cabinet-type hose washing machine washes, rinses, and drains fire hose (Figure 2.9). This automated washer, which can be operated by one person, can be used with or without detergent. A jet-spray washer can be attached to a section of hose or attached directly to a hydrant (Figure 2.10). Water under high pressure is directed

Figure 2.10 A jet spray washer cleans the hose jacket with a high pressure water stream that surrounds the hose.

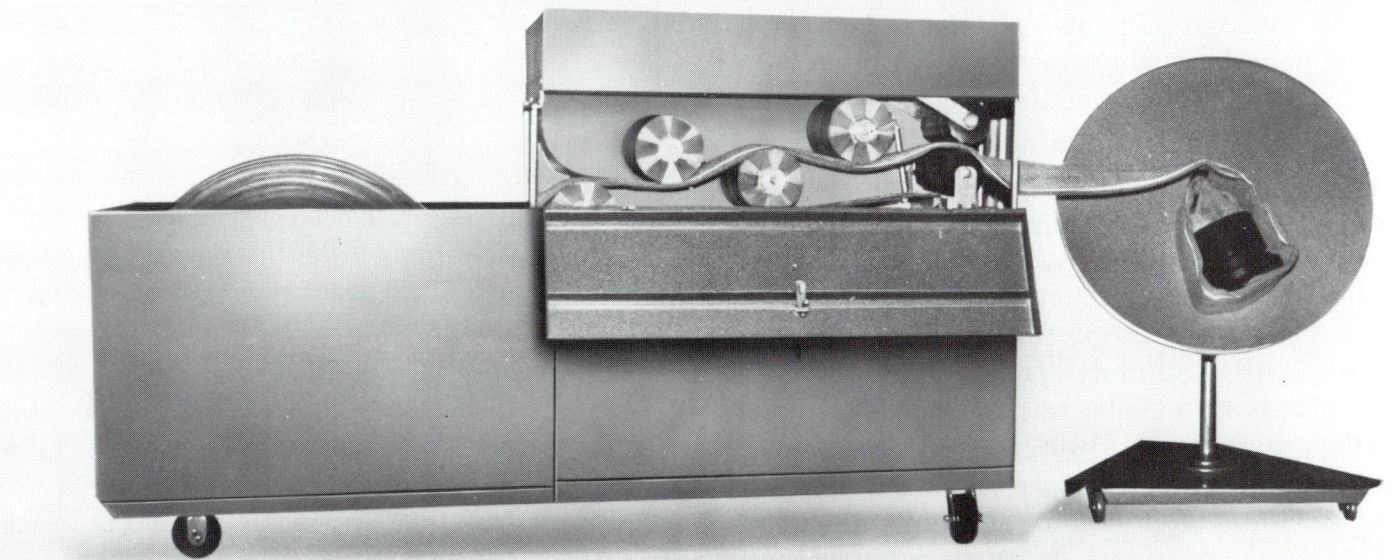

Figure 2.9 A cabinet style hose washer. *Courtesy of Circul-Air, Inc.*

through small jets within the washer housing so that as hose is guided through the washer, dirt and debris are removed.

If these washing devices are not available, an adequate job of cleaning can be done by hand. Simply brushing off accumulations of dirt, leaves, and other debris with dry brooms, however, does not adequately clean hose. Although this practice leaves the hose looking clean, tiny particles of grit and sand remain within the jacket among the threads, where they cause wear to individual fibers. To remove this grit, wash hose with clear water as soon as possible after use. Scrub the hose jacket with brushes or brooms along with a high-pressure water stream (Figure 2.11). Use a mild

Figure 2.12 Clean coupling threads thoroughly with a stiff bristle brush.

Figure 2.11 Use a high pressure water stream and a brush or broom to scrub hose clean.

soap solution if the hose has been exposed to oil or chemicals. This helps break the chemical bonds between the contaminants and the jacket fibers. Rinse hose thoroughly after a soap washing to remove all traces of contaminants and soap.

When a coupling swivel becomes stiff or sluggish with dirt or other foreign matter, remove the gasket and put the coupling in a container of warm, soapy water. Work the swivel back and forth to help break loose accumulations of dirt between the swivel collar and coupling body. Clean coupling threads with a stiff bristle brush (Figure 2.12). If the threads are occluded with tar, asphalt, or other foreign matter, use a wire brush to break loose stubborn material. Rinse the coupling thoroughly in clear water after washing and lubricate according to manufacturer's instructions. Lubricants such as graphite or silicone are usually all that is needed to maintain swivels so that they spin freely. If the gasket is cracked or scored (Figure 2.13), or if it has become hardened and inflexible, replace it.

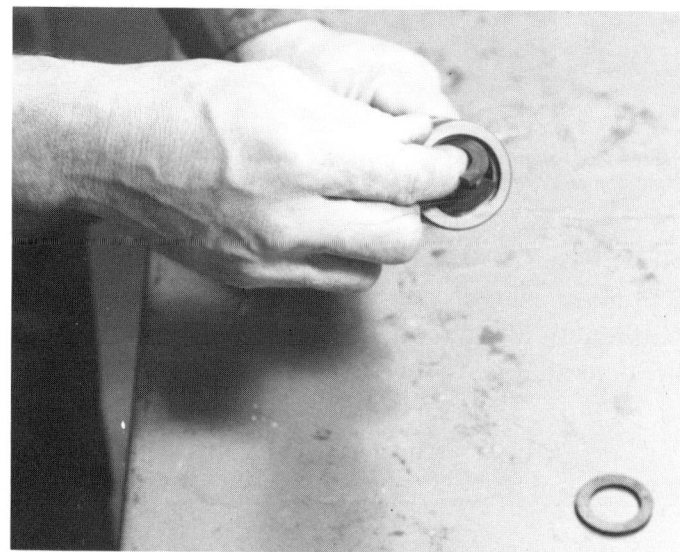

Figure 2.13 The gasket inside the swivel should be replaced if defective.

After fire hose has been thoroughly washed, it should be either hung to dry in a hose tower, placed on an inclined drying rack, or dried in a cabinet-type hose dryer (Figure 2.14). An alternate drying method is to lay hose on edge on the floor of a large room such as an apparatus room (Figure 2.15). The disadvantage of drying hose in this manner is that it is difficult to completely drain residual water from within the hose.

Hose towers and drying racks should be adequately ventilated and protected so that hose is

Care, Maintenance, and Testing **39**

Hose dryers may be purchased commercially or may be built into a building. Dryers should be provided with adequate fans for forced air circulation within the drying space and with thermostats to control interior temperatures. Moisture from the drying hose should be vented to the outside. Racks should be removable to allow easy placement and arrangement of hose sections for thorough drying. A timing device should also be provided to prevent the hose from being left in the dryer too long, which could result in damage from prolonged heat exposure.

Figure 2.14 A cabinet-type hose dryer. *Courtesy of Circul-Air, Inc.*

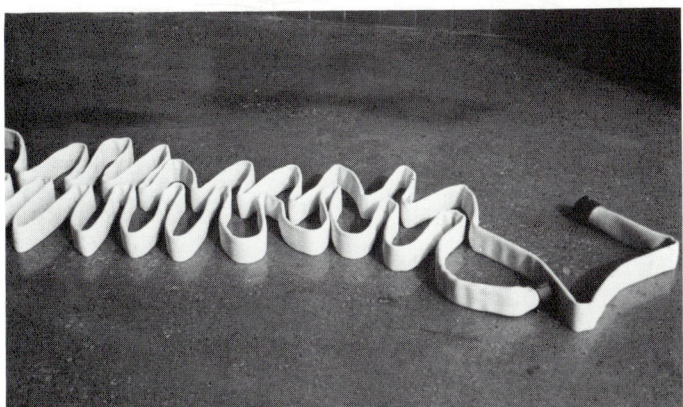

Figure 2.15 Hose can be laid out to dry in an apparatus room.

not exposed to excessive temperatures or direct sunlight (Figure 2.16). Hose in outside drying towers should be secured so that it cannot swing in a wind. Such movement could cause couplings to bang against each other or against tower supports, resulting in coupling damage. Drying racks should be inclined enough to promote water to drain from the hose during drying. Avoid placing hose sections too closely together or to touch, which slows the drying process.

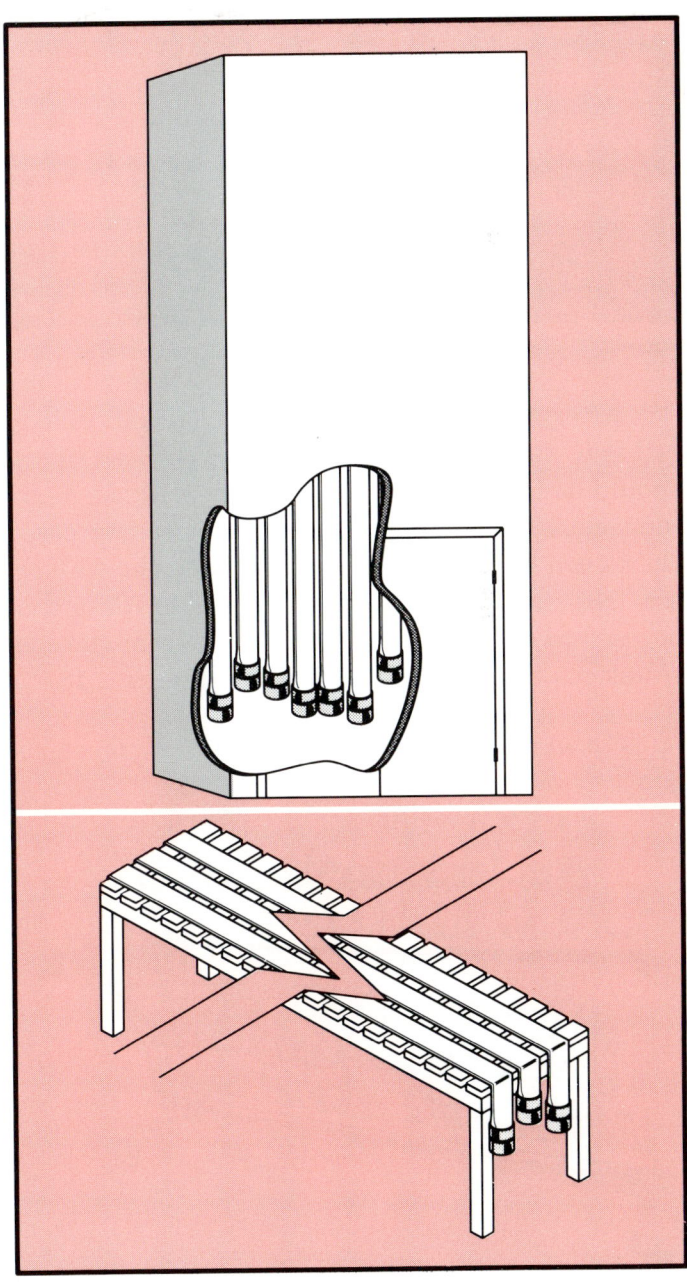

Figure 2.16 Hose towers and drying racks should have adequate ventilation and should protect hose from direct sunlight.

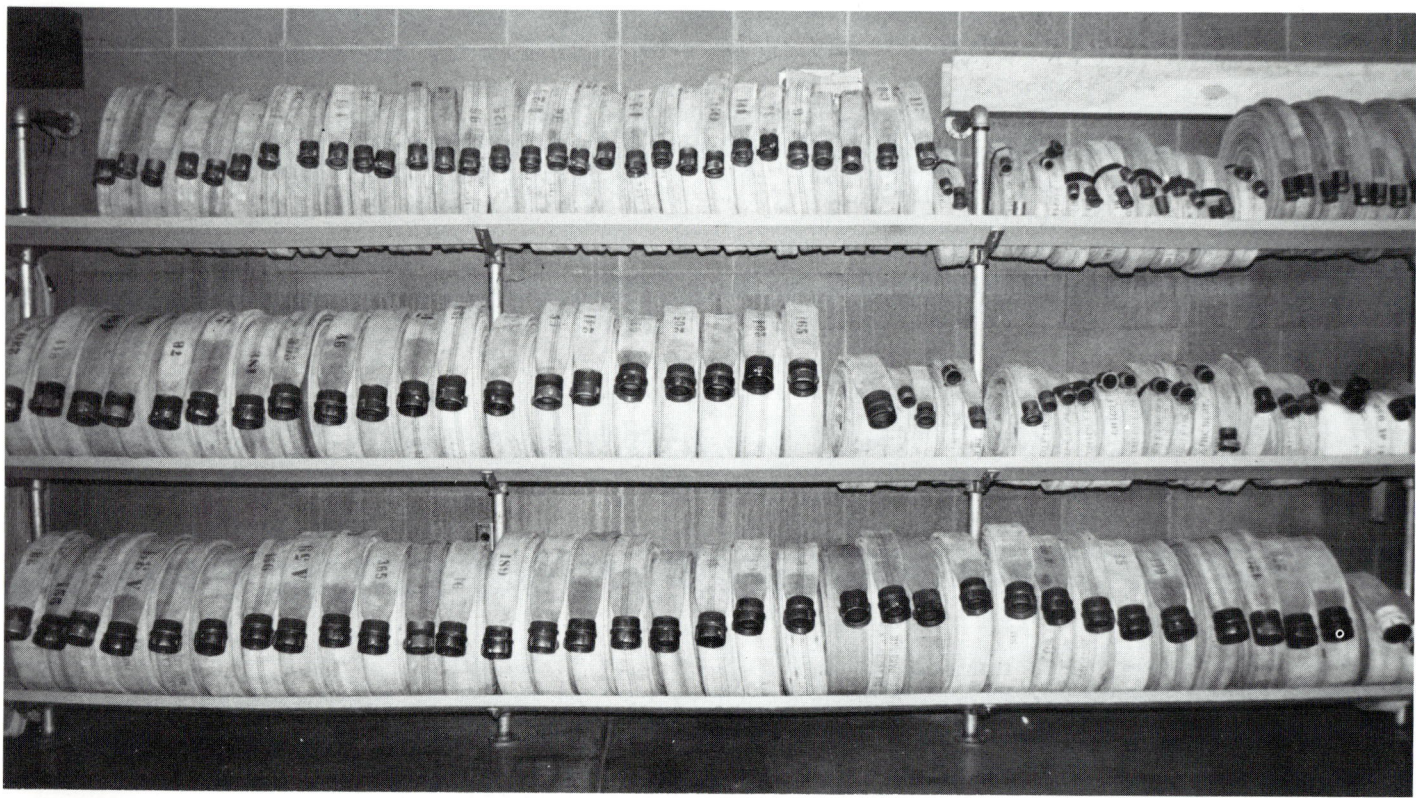

Figure 2.17 Clean, dry hose should be rolled and stored on racks.

After fire hose has been washed and dried, roll and store it in racks (Figure 2.17). Hose racks should be located in a clean, well-ventilated room in or close to the apparatus room for easy access. Racks can be mounted permanently on the wall or can free-stand on the floor. Mobile hose racks can be used to both store and move hose from storage rooms out to apparatus for loading (Figure 2.18).

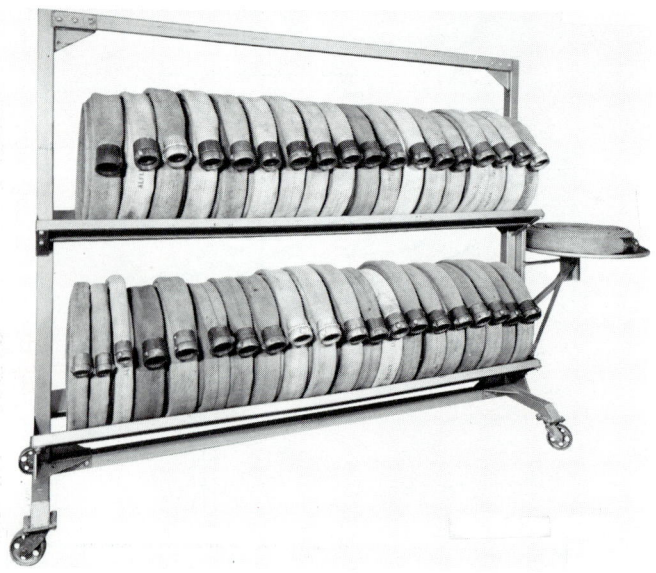

Figure 2.18 A mobile hose rack. *Courtesy of Circul-Air, Inc.*

Inspection

Each time that a section of hose is used, whether for fire fighting or for training, it should be inspected to ensure that it is free of visible soiling or damage. If any of the following deficiencies are found, they should be corrected before the hose is stored or placed back in service:

- Dirt or debris on the hose jacket or couplings
- Damage to the hose jacket
- Evidence that the coupling is coming loose from the hose
- Damage to male and female threads
- Obstructed operation of the swivel
- Absence of a well-fitting gasket in the swivel

HOW TO RECOUPLE FIRE HOSE

Most damage to fire hose is repairable and these repairs can be made by fire department personnel. Some repair is very simple, such as previously described for threads and couplings. More complex repair, such as recoupling, requires special tools and training, but can still be easily accomplished in the fire department shop.

If a hose bursts near a coupling, the portion of hose between the hole and the coupling can be removed and the remaining hose, slightly shortened, can be recoupled. If a hole appears too far away from the coupling, however, the hose must be taken out of service because most kinds of fire hose cannot be patched. The determining factor in deciding whether to recouple a hose or take it out of service is usually a matter of agency policy. An example of such a rule is: "If fire fighting hose requires recoupling, it shall not be shortened to less than 90 percent of its original length."

Each procedure described below requires that the hose end be cut squarely so that it fits tightly to the coupling. Use a straight edge hose cutter or similar device to square cut the frayed edge of the hose before recoupling (Figure 2.19).

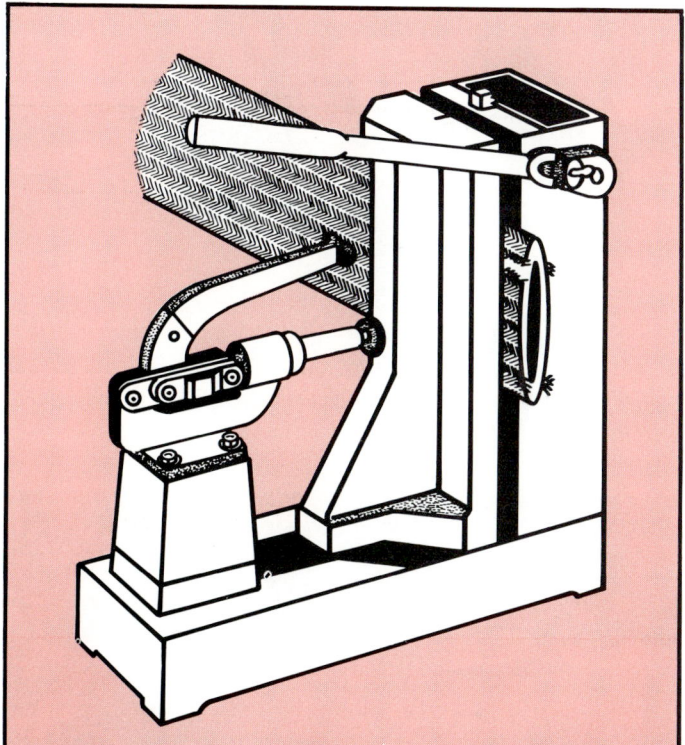

Figure 2.19 A straight edge hose cutter.

Couplings Attached with Expansion Rings

Couplings designed for attachment with expansion rings are manufactured with serrations on the inside of the shank. The procedure involves placing an expansion ring, which is slightly smaller than the hose, inside the end of the hose. The hose and ring are pushed into the coupling shank against a gasket, then the ring is expanded against the hose to compress it against the coupling shank, thus affixing it permanently to the hose.

The procedure requires a machine called an expander. There are three types of expanders: manually operated, hand-hydraulic, and power (Figure 2.20). All expanders contain a mandrel assembly, which fits inside the expansion ring and

Figure 2.20 From top to bottom, a manual expander, hand-hydraulic expander, and power expander. *Courtesy of Niedner, Ltd.; Bell & Pell Company, Inc.; and Akron Brass.*

spreads to expand the ring against the hose and coupling. Each size coupling requires a different size mandrel (Figure 2.21).

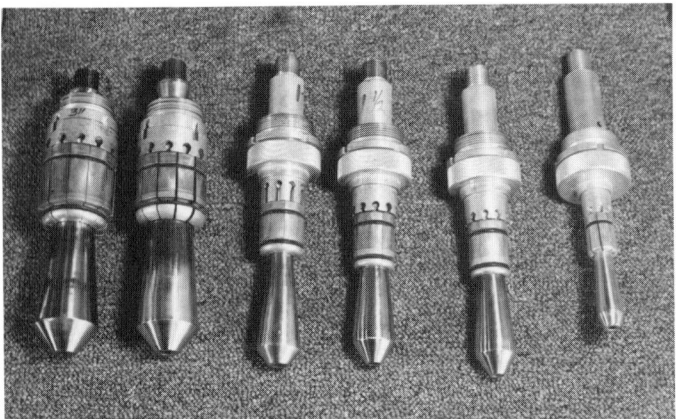

Figure 2.21 Mandrels come in sizes to fit each hose diameter.

Expansion rings vary not only in diameter, but also in length (Figure 2.22). Because fire service hose is made to withstand a higher internal pressure than non-fire service hose, it requires a longer expansion ring. This gives it more surface area to bind the hose to the coupling shank. It is important, therefore, that the proper type of expansion rings are purchased and used for recoupling fire service hose.

Figure 2.22 Expansion rings vary in diameter and length. *Courtesy of Niedner, Ltd.*

Figure 2.23 An expansion ring gasket (right) has a smaller surface area than a standard coupling gasket (left). *Courtesy of Niedner, Ltd.*

Expansion ring gaskets, which prevent seepage around the end of the hose, fit between the end of the hose and the coupling. Although similar to a standard coupling gasket, the expansion ring gasket has a smaller surface area (Figure 2.23).

The procedure for attaching a female expansion ring coupling to a hose is as follows:

Step 1: Install the segment assembly into the nosepiece of the machine (Figure 2.24). The tapered drawbar threads, as well as the segment holder, must be bottomed.

Step 2: Select the suggested pressure from the pressure chart and preset the designated pressure by using the pressure regulator knob (Figure 2.25). Pressure requirements vary according to the type of hose, as well as to whether the couplings are made of brass or aluminum.

Figure 2.24 STEP 1: Install the segment assembly into the nosepiece. *Courtesy of Niedner, Ltd.*

Figure 2.25 STEP 2: Adjust the pressure regulator knob. *Courtesy of Niedner, Ltd.*

Step 3: Measure the length of the coupling waterway (Figure 2.26), then add 1/16 of an inch (1.6 mm) to this measurement.

Step 4: Rotate the adjusting collar to set the measured distance on the machine (Figure 2.27). (Measure to the *back* of the adjusting collar for female couplings; measure to the *lip* of the adjusting collar for male couplings).

Step 5: Place a coupling over the segment assembly so that the coupling face fits squarely against the adjusting collar. Expand the segments against the inside

Figure 2.26 STEP 3: Measure the length of the waterway. *Courtesy of Niedner, Ltd.*

Figure 2.27 STEP 4: Rotate the adjusting collar to set the measured distance. *Courtesy of Niedner, Ltd.*

of the coupling to check that the measurement is correct. The segments should contact only the inside of the bowl and not touch the waterway (Figure 2.28).

Step 6: Remove the coupling, then place the expansion ring over the segment assembly so that the ring sets flush against the segment holder (Figure 2.29).

Step 7: Insert a backup gasket into the coupling, then push the hose into the coupling bowl (Figure 2.30). Make sure that the hose end is tight against the gasket.

Step 8: Place the hose and coupling over the segment assembly and expansion ring so that the coupling face fits squarely against the adjusting collar (Figure 2.31).

Figure 2.28 STEP 5: Place a coupling (with or without hose) over the segment assembly to check the measurement. Expand the segments against the inside of the coupling. The segments should not touch the waterway. *Courtesy of Niedner, Ltd.*

Figure 2.29 STEP 6: Remove the coupling, then place the expansion ring over the segment assembly and against the segment holder. *Courtesy of Niedner, Ltd.*

Figure 2.30 STEP 7: Push the hose into the coupling bowl against the backup gasket. *Courtesy of Niedner, Ltd.*

Figure 2.31 STEP 8: Push the coupling over the expansion ring and segment assembly. *Courtesy of Niedner, Ltd.*

Step 9: Place the directional handle in the EXPAND position and pump until the desired pressure is reached (Figure 2.32).

Step 10: Place the directional handle in the RETRACT position and back off the mandrel until you can rotate the coupling. Rotate the coupling a quarter-turn (Figure 2.33). Place the directional handle in the EXPAND position and pump again to the desired pressure. This will ensure a snug, tight fit.

Step 11: Place the directional handle in the RETRACT position and release mandrel pressure until you can remove the coupled hose (Figure 2.34).

Figure 2.32 STEP 9: Place the directional handle in the EXPAND position and pump until the desired pressure is reached. *Courtesy of Niedner, Ltd.*

Figure 2.33 STEP 10: Release mandrel pressure until you can rotate the coupling a quarter-turn, then repeat Step 9. *Courtesy of Niedner, Ltd.*

Figure 2.34 STEP 11: Remove the coupled hose from the mandrel.

After the coupling has been attached, inspect the expansion ring to be sure it is slightly indented with the outline of segments and that the rubber behind the expansion ring indicates some rolling effect due to compression. Check the waterway of the coupling to be sure that the segments did not indent the waterway. If indenting is visible, the locating plate was not adjusted properly. Look for hairline fractures, a rippling effect, or distortion on the outside of the coupling. Such damage indicates that too high a pressure was used.

After the inspection process, test the recoupled hose section at the recommended service test pressure to confirm that a satisfactory attachment has been made.

Couplings Attached with Screw-In Expanders

Hose subjected to high pressures, such as booster hose, requires a coupling that is attached very securely. A coupling with a screw-in expander is capable of withstanding extremely high pressure because of its design. As illustrated in Figure 2.35, the screw-in expander is a separate but integral component of the coupling body. It is threaded in two ways: it has a fine thread along most of its length so that it may be screwed into the coupling shell, and it has standard threads so that the complete coupling assembly can be connected to another coupling.

Figure 2.35 A screw-in expander. *Courtesy of Weis American Fire Equipment, Oklahoma City, OK.*

The procedure for attaching this type of coupling is as follows:

MALE COUPLING PROCEDURE

Step 1: Lubricate the shell and hose end with a soapy water solution or lubricant (Figure 2.36). Use an inert oil-based or silicone-based lubricant, as recommended by the coupling manufacturer.

Step 2: Screw the end of the hose into the shell until the hose bottoms in the shell (Figure 2.37).

Step 3: Scribe the hose at the end of the shell with indelible ink (Figure 2.38). This mark will be used to check later for slippage.

Step 4: Place the clamps around the shell (Figure 2.39).

Step 5: Put the clamps, shell, and hose assembly into a vise; tighten the vise jaws against the clamps until the shell cannot turn (Figure 2.40).

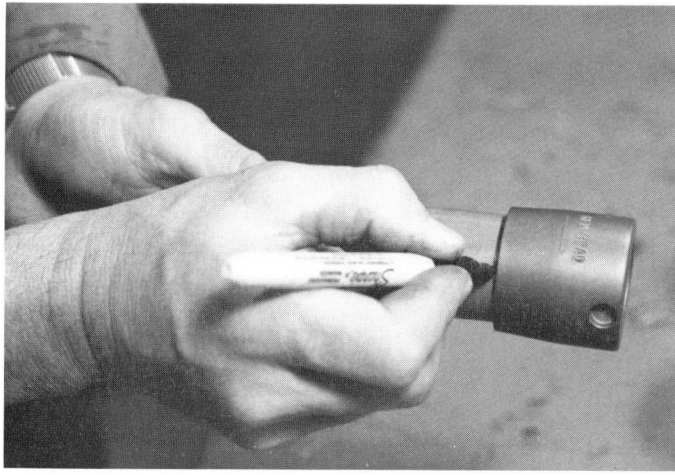

Figure 2.38 STEP 3: Mark the hose with indelible ink. *Courtesy of Weis American Fire Equipment, Oklahoma City, OK.*

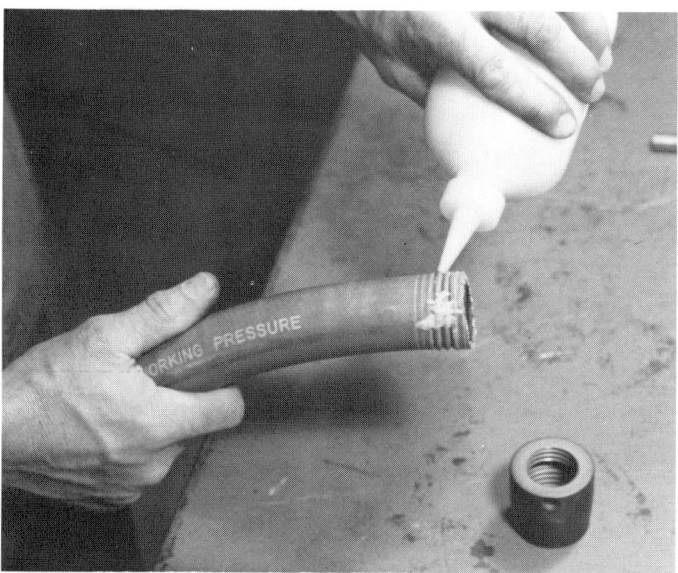

Figure 2.36 STEP 1: Lubricate the shell and hose end. *Courtesy of Weis American Fire Equipment, Oklahoma City, OK.*

Figure 2.39 STEP 4: Place the clamps around the shell. *Courtesy of Weis American Fire Equipment, Oklahoma City, OK.*

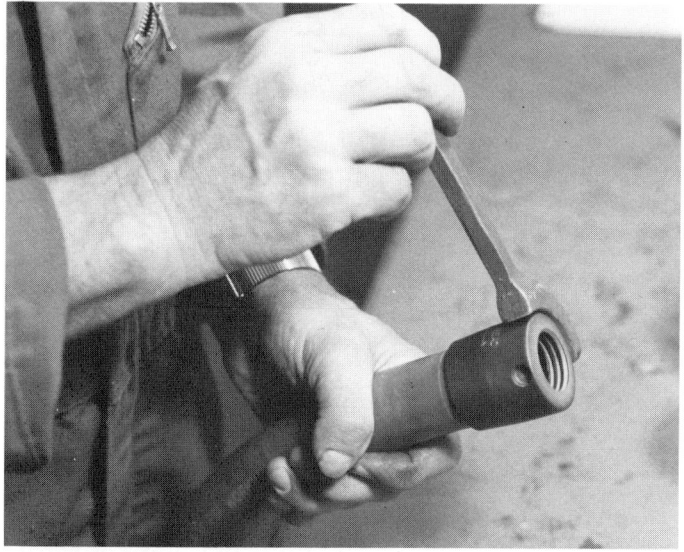

Figure 2.37 STEP 2: Screw the hose into the shell until it bottoms. *Courtesy of Weis American Fire Equipment, Oklahoma City, OK.*

Figure 2.40 STEP 5: Tighten the vise jaws against the clamps until the shell cannot turn. *Courtesy of Weis American Fire Equipment, Oklahoma City, OK.*

Step 6: Place the expander on the key so that the key engages the interior grooves and protrudes from the leading end of the expander (Figure 2.41). This protrusion serves as an entrance guide for the expander.

Step 7: Lubricate the threaded exterior of the expander and the interior of the hose end (Figure 2.42).

Step 8: Screw the expander into the shell and hose until the expander bottoms within the shell (Figure 2.43). When driving the expander into the hose and shell,

- Do *not* interrupt the rotations, once started.
- Do *not* permit the hose to rotate.
- Do *not* allow the key to push up.

Step 9: Release tension on the vise, remove the clamps, and inspect the coupling for slippage. The coupling should not have moved away from the scribe mark.

Figure 2.41 STEP 6: Place the expander on the key. *Courtesy of Weis American Fire Equipment, Oklahoma City, OK.*

Figure 2.42 STEP 7: Lubricate the interior of the hose end and the threaded exterior of the expander body. *Courtesy of Weis American Fire Equipment, Oklahoma City, OK.*

Figure 2.43 STEP 8: Screw the expander into the shell and hose. *Courtesy of Weis American Fire Equipment, Oklahoma City, OK.*

The procedure for attaching the female coupling is essentially the same as for the male coupling, except that the expander has a female-threaded swivel nut rather than male threads. Screw the expander into the shell and hose until the expander bottoms within the shell, then back the expander off until the swivel nut turns freely. The final step in the female coupling attachment process is to place a gasket in the swivel (Figure 2.44).

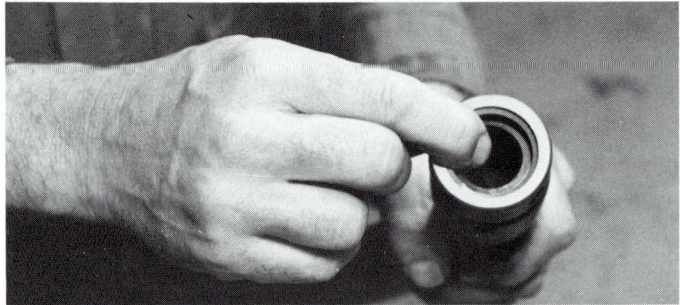

Figure 2.44 Place a gasket in the swivel. *Courtesy of Weis American Fire Equipment, Oklahoma City, Ok.*

Couplings Attached with Bolted-On Collars

Couplings with bolted-on collars require only an allen wrench for attachment. The steps for attaching a collared coupling are as follows:

Step 1: Slide the hose over the coupling shank (Figure 2.45).

Step 2: Fit the collar over the hose and the shank (Figure 2.46).

Step 3: Insert the bolts and tighten with approximately 40 foot pounds (54 N·m) of torque (Figure 2.47).

Care, Maintenance, and Testing **47**

Figure 2.45 STEP 1: Slide the hose over the coupling shank. *Courtesy of Niedner, Ltd.*

Figure 2.46 STEP 2: Fit the collar over the hose and the shank. *Courtesy of Niedner, Ltd.*

Figure 2.47 STEP 3: Insert and tighten the bolts. This is commonly done with an Allen wrench. *Courtesy of Niedner, Ltd.*

Couplings Attached with Tension Rings

Couplings attached with tension rings (also known as "contractual sleeve couplings") have several components (Figure 2.48): coupling body, flange ring, tension ring, and clamp ring (contractual sleeve). These couplings require a special tool kit to install. They also require a special disassembly key to remove them from the hose, because prying the components apart with screwdrivers or

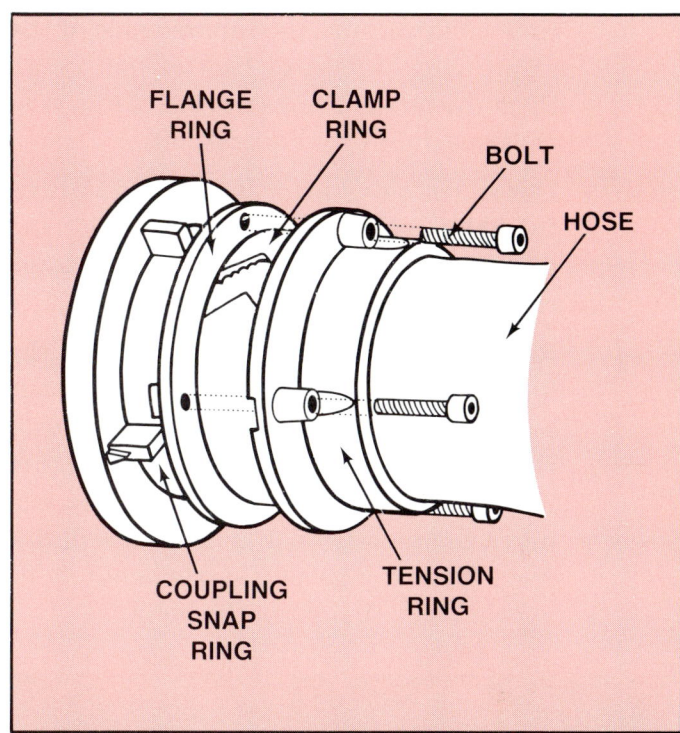

Figure 2.48 A tension ring coupling is made of several components.

other similar tools could result in damage. The steps for attaching a tension ring coupling are as follows:

Step 1: Slip the tension ring and flange ring over the end of the hose, then slide the hose onto the coupling shank so that it butts against the coupling snap ring (Figure 2.49).

Step 2: Move the flange ring up against the snap ring, then place the lubricated nylon clamp ring over the hose behind the flange

Figure 2.49 STEP 1: Slip the tension ring and flange ring over the end of the hose, then slide the hose onto the coupling shank so that it butts against the coupling snap ring. *Courtesy of Snaptite, Inc.*

ring (making sure the tapered side of the clamp ring is toward the tension ring). Push the tension ring onto the clamp ring as far as it will go (Figure 2.50).

Step 3: Install the two starter bolts, then tighten the bolts evenly to draw the tension ring up to within ¼-inch (6 mm) of the flange ring (Figure 2.51).

Step 4: Remove the starter bolts and install the four tension ring bolts. Tighten the bolts evenly to draw the tension ring as far as it will go toward the flange ring (Figure 2.52). **NOTE:** The tension ring may not move completely flush against the flange ring. If the flange ring starts bending, do *not* continue to tighten the bolts.

Figure 2.50 STEP 2: Move the flange ring up against the snap ring, then place the lubricated nylon clamp ring over the hose behind the flange ring (making sure the tapered side of the clamp ring is toward the tension ring). Push the tension ring onto the clamp ring as far as it will go. *Courtesy of Snaptite, Inc.*

Figure 2.51 STEP 3: Install the two starter bolts, then tighten the bolts evenly to draw the tension ring up to within ¼-inch (6 mm) of the flange ring. *Courtesy of Snaptite, Inc.*

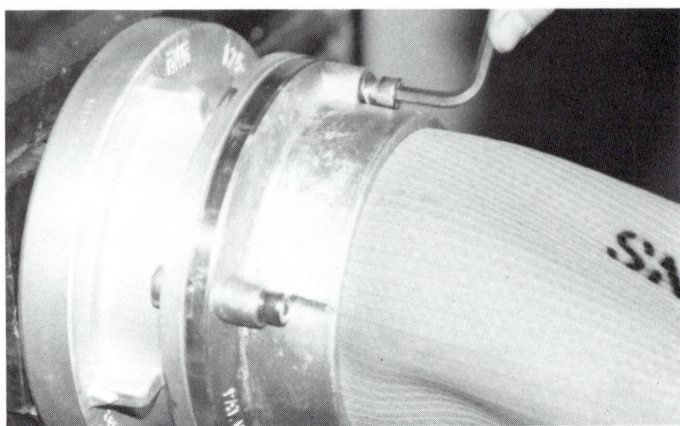
Figure 2.52 STEP 4: Remove the starter bolts and install the four tension ring bolts. Tighten the bolts evenly to draw the tension ring further toward the flange ring. *Courtesy of Snaptite, Inc.*

HOW TO SERVICE TEST HOSE

There are two types of tests for fire hose: acceptance testing and service testing. At the request of the purchasing agency, coupled hose is acceptance tested by the manufacturer before the hose is shipped. This type of testing is relatively rigorous, and the hose is subjected to extremely high pressures to ensure that it can withstand the most extreme conditions in the field. This type of testing should *not* be attempted by the fire department. Service testing is done periodically by the user to ensure that the hose is being maintained in optimum condition. This testing of in-service hose confirms that it is still able to function under maximum pressure during fire fighting or other operations. Guidelines for both types of tests are in the latest NFPA 1962, *Standard for the Care, Use and Maintenance of Fire Hose Including Connections and Nozzles.*

Fire department hose should be service tested annually. Unlined standpipe hose should be tested five years from the date of purchase, again at the eighth year, and every two years thereafter.

Before performing a service test the hose should be examined for jacket defects, coupling damage, and worn or defective gaskets. Defects should be corrected, if possible, prior to testing. If damage is not repairable, the hose should be taken out of service.

Test Site Preparation

Hose should be tested in a place that has adequate room to lay out the hose in straight runs, free of kinks or twists. The site should be isolated from traffic and, if testing is done at night, be well lighted. Its surface should be smooth and free of dirt and debris. A water source sufficient for filling the hose is also necessary.

The following equipment is needed to service test hose:

- A hose testing machine (Figure 2.53), portable pump, or fire department pumper. All such equipment should be equipped with gauges certified as accurate within one year prior to testing.
- A hose test gate valve.
- A means of recording the hose numbers and test results.
- Tags or other means to identify sections that fail.
- Nozzles with shutoff valves.

Figure 2.53 A hose testing machine. *Courtesy of Rice Hydro Equipment Manufacturing.*

Safety at the Hose Testing Site

As when working with any equipment, exercise care when working with hose, especially when it is under pressure. Air is compressible, and the sudden release of expanding air when a pressurized hose bursts can result in a serious injury or fatal accident. Pressurized hose is potentially dangerous because of its tendency to whip back and forth if a break occurs, such as when a coupling pulls loose. To prevent this, use a specially designed hose test gate valve (Figure 2.54). This is a valve with a ¼-inch (6 mm) hole in the gate that

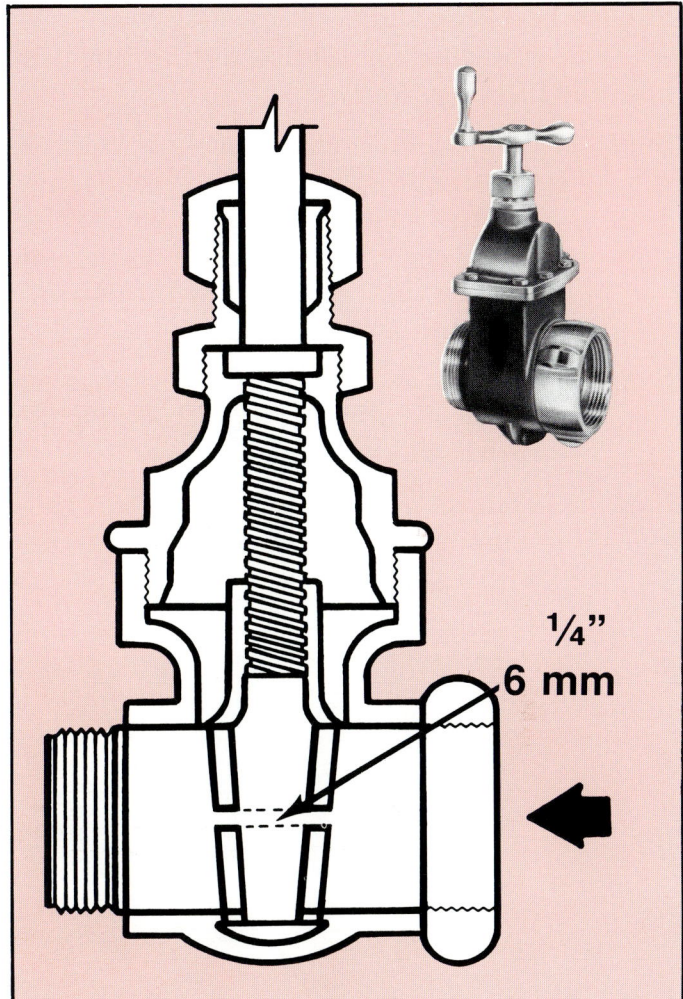

Figure 2.54 A hose test gate valve.

permits pressurizing the hose but will not allow water to surge through the hose if the hose fails. Even when using the test valve, stand or walk near the pressurized hose only as necessary.

Open and close all valves slowly to prevent water hammer in the hose and pump. Test lengths of hose should not exceed 300 feet (91 m) in length (longer lengths are more difficult to purge of air).

Lay LDH flat on the ground prior to charging. This will prevent unnecessary wear at the edges. Stand away from the LDH discharge valve connection when charging because some types tend to twist when filled with water and pressurized. This could cause the connection to twist loose.

Keep the hose testing area free of water when filling and discharging air from the hoses. This will aid in detecting minor leaks around couplings during testing.

50 HOSE

Service Test Procedure

The procedure for service testing lined fire hose and LDH is as follows:

Step 1: Connect a number of hose sections (check the gaskets before connecting) into test lengths of no more than 300 feet (91 m) each. Tighten the connections between the sections with spanner wrenches (Figure 2.55).

Step 2: Connect an *open* test valve to each discharge valve used (Figure 2.56). Tighten each connection with spanners.

Step 3: Connect a test length to each test valve (Figure 2.57). Tighten each connection with a spanner.

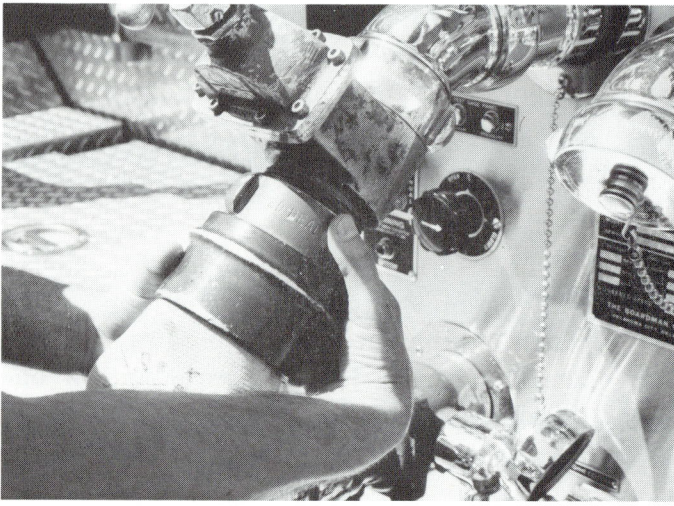

Figure 2.57 STEP 3: Connect a test length to each test valve.

Step 4: Tie a rope, rope hose tool, or hose strap to each test length of hose 10 to 15 inches (254 mm to 380 mm) from the test valve connections. Secure the other end to the discharge pipe or other nearby anchor (Figure 2.58).

Step 5: Attach a shutoff nozzle (or any device that will permit water and air to drain from the hose) to the open end of each test length (Figure 2.59).

Step 6: Fill each hoseline with water with a pump pressure of 50 psi (345 kPa). Open the nozzles as the hoselines are filling. Hold them above the level of the pump discharge to permit all the air in the hose to discharge (Figure 2.60). Discharge the water away from the test area.

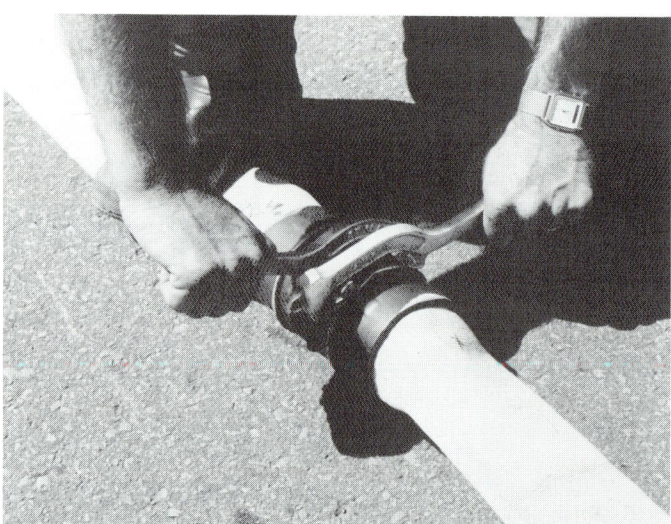

Figure 2.55 STEP 1: Connect the hose together and tighten all connections with a spanner wrench.

Figure 2.56 STEP 2: Connect an open test valve to each discharge valve used.

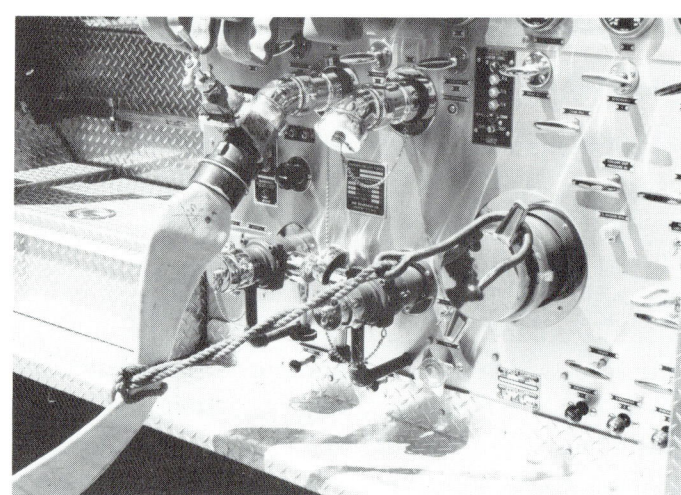

Figure 2.58 STEP 4: Secure the hose with a strap or rope.

Care, Maintenance, and Testing **51**

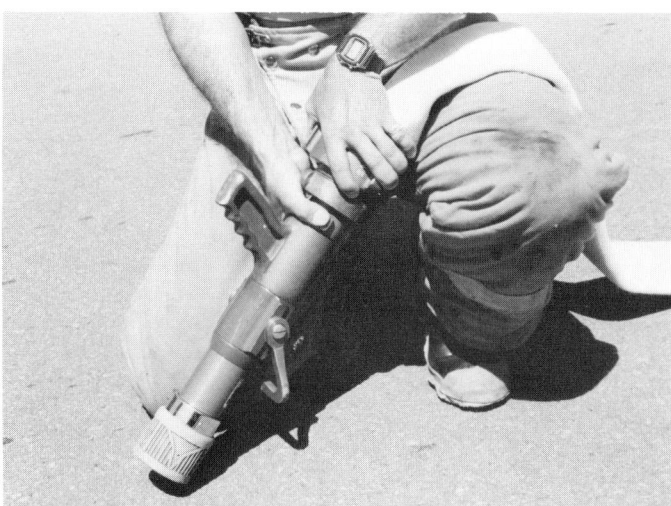

Figure 2.59 STEP 5: Attach a nozzle or shutoff valve to each test length.

Figure 2.60 STEP 6: Discharge the air from each test length.

Step 7: Close the nozzles after all air has been purged from each test length. Make a chalk or pencil mark on the hose jackets against each coupling (Figure 2.61). Check that all hose is free of kinks and twists.

Figure 2.61 STEP 7: Mark the hose with a pencil mark or chalkline against each coupling.

Step 8: Close each hose test gate valve (Figure 2.62).

Step 9: Increase the pump pressure to the required test pressure (Figure 2.63) while personnel closely monitor the connections for leakage as the pressure increases.

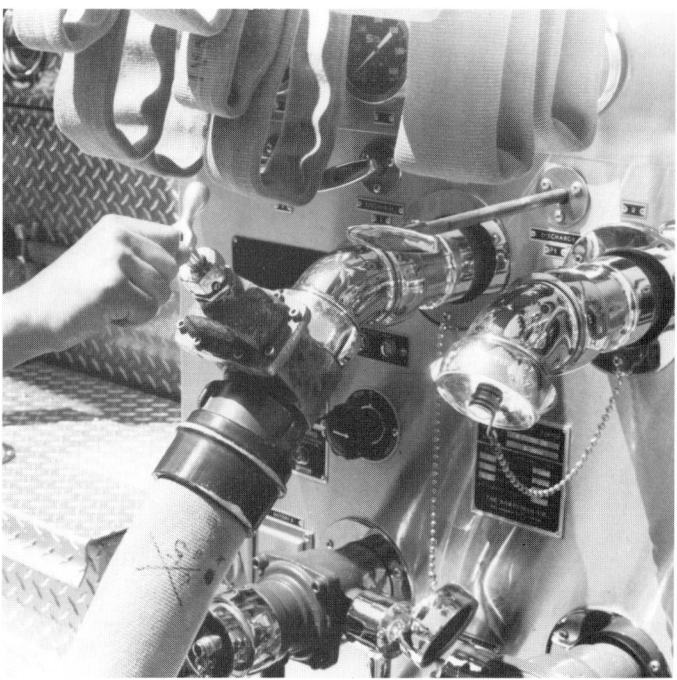

Figure 2.62 STEP 8: Close each hose test gate valve.

Figure 2.63 STEP 9: Increase the pump pressure to the required test pressure.

Step 10: Maintain the test pressure for five minutes. Inspect all couplings to check for leakage ("weeping") at the point of attachment (Figure 2.64).

Step 11: After five minutes, slowly reduce the pump pressure, close each discharge valve (Figure 2.65), and disengage the pump.

Step 12: Slowly open each nozzle to bleed off pressure in the test lengths. Break all the hose connections and drain water from the test area (Figure 2.66).

Step 13: Observe marks placed on the hose at the couplings. If a coupling has moved during the test, tag the hose section for recoupling. Tag all hose that has leaked or failed in any other way (Figure 2.67).

NOTE: Expect a 1/16- to ⅛-inch (1.6 mm to 3.2 mm) uniform movement of the coupling on newly coupled hose. This slippage is normal during initial testing but should not occur during subsequent tests.

Step 14: Record the test results for each section of hose.

Figure 2.65 STEP 11: After five minutes, reduce the pump pressure, close each discharge valve, and disengage the pump.

Figure 2.66 STEP 12: Break the hose connections and drain water away from the test area.

Figure 2.64 STEP 10: Inspect all couplings for leaks.

Figure 2.67 STEP 13: Tag leaking or failed hose for recoupling.

Testing Unlined Linen Hose

The procedure for testing unlined linen hose is the same as for lined hose, with a number of exceptions. Because this type of hose is unlined, it leaks under pressure. It should be expected, therefore, that some water flow be maintained to compensate for leakage through the jacket. For this reason, the hose test gate valve cannot be used when testing unlined hose.

Fill the hose with water at a 50 psi (345 kPa) pump pressure. Open the nozzle(s) to remove the air, then close the nozzle(s); maintain the 50 psi (345 kPa) pressure for 10 minutes to allow the hose to wet-soak. This causes the linen yarn to swell, which reduces leakage.

Increase the pump pressure to the required test pressure. The hose should leak no more than 20 gpm (76 L/min). Use a flowmeter to monitor this water flow. If the test pressure cannot be maintained without exceeding this flow rate, remove the hose from service.

Record the test results for each section of hose. Dry the hose immediately to prevent mildew.

Vacuum Testing Hard Suction Hose

There are two reasons for vacuum testing hard suction hose: to check for leaks, and to check for a loose lining that may protrude into the waterway and restrict water flow. The following equipment, in addition to that listed for testing lined hose, is needed to vacuum-test hard suction hose:

- Vacuum pump, if the above equipment is incapable of pulling a vacuum
- Vacuum gauge
- Transparent plastic disk, large enough to cover the end of the hard suction hose
- Flashlight

The procedure for vacuum-testing a hard suction hose is as follows:

Step 1: Close the valve between the tank and pump, then completely drain the pump (Figure 2.68). Cap and tighten all discharge and intake openings to prevent a vacuum leak.

Step 2: Place a lighted flashlight inside the barrel of the pump intake with the light directed outward (Figure 2.69). (The intake barrel should have an internal screen.)

Figure 2.68 STEP 1: Close the valve between the tank and pump, then completely drain the pump. Cap and tighten all discharge and intake openings to prevent a vacuum leak.

Figure 2.69 STEP 2: Place a flashlight inside the barrel with the light shining outward.

Step 3: Connect the hard suction hose to the intake and provide support so that the hose is held horizontal from the pump intake (Figure 2.70).

Step 4: Place the transparent disk over the male end of the hose (Figure 2.71). A small amount of heavy lubricating grease applied to the edge of the hard suction coupling will help seal the disk to the coupling when pulling a vacuum.

Figure 2.70 STEP 3: Connect the hard suction and support the hose so that it is horizontal to the pump intake.

Figure 2.71 STEP 4: Place the transparent disk over the male end of the hose.

Step 5: Prime the pump until at least 22 inches (559 mm) of mercury reads on the vacuum gauge (Figure 2.72).

Step 6: Discontinue priming, then shut down the engine and listen for vacuum leaks (Figure 2.73). Monitor the vacuum gauge. No more than 10 inches (254 mm) of vacuum should be lost within 10 minutes.

NOTE: If the vacuum loss exceeds 10 inches/10 minutes (254 mm/10 minutes) and the source of the leak cannot be identified, disconnect the hard suction hose and remove the flashlight, replace the intake cap, and repeat the priming procedure. If the vacuum loss reoccurs, the leak may be in the pump.

Figure 2.72 STEP 5: Prime the pump until at least 22 inches (559 mm) of mercury reads on the vacuum gauge.

Figure 2.73 STEP 6: Discontinue priming, then shut down the engine and listen for vacuum leaks.

Care, Maintenance, and Testing 55

Step 7: Look through the transparent disk into the hose (Figure 2.74). Examine the lining for any indication that it is coming loose and protruding into the waterway.

Step 8: After 10 minutes, open the pump drain slowly to allow air to reenter the pump. Disconnect the hard suction hose and remove the flashlight. Replace the intake cap and prime the pump to prepare it for normal service.

Step 9: Record the test results for each section of hard suction hose.

mark each section of hose. Mark hose by using one or more of the following methods:

- Die-stamp the section identification number and fire department initials on a coupling (Figure 2.75).
- Stencil the identification number and fire department initials on the hose jacket near each end using indelible ink (Figure 2.76).
- Color-code the coupling shank, swivel, or lugs to indicate the apparatus or fire station to which the hose is permanently assigned.

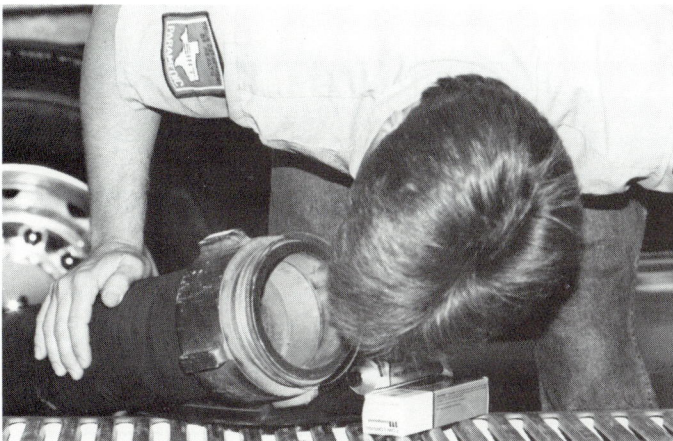
Figure 2.74 STEP 7: Look through the transparent disk to inspect the lining.

PATCHING HOSE

It is strongly recommended that fire hose not be patched. Some types of covered and extruded hose are repairable by patching, but this should be restricted to the repair of surface blemishes. There are a number of methods for patching hose, each of which varies according to the manufacturer's recommendations.

Some LDH can be repaired by using a hose splice. This is done by cutting the hose on either side of the damage to remove it, and attaching the free ends of the hose to a special in-line splice with coupling bindings. Again, this procedure will vary by manufacturer, and is not recommended for all types of LDH.

RECORDS OF HOSE TESTS AND INSPECTIONS

To keep required records of hose tests, repairs, and an inventory, it is necessary to permanently

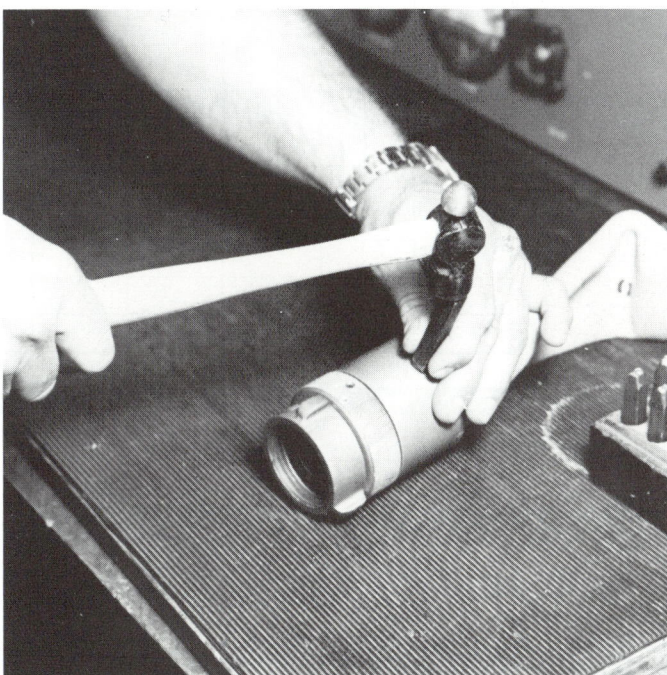
Figure 2.75 Die-stamp the section identification number on the coupling.

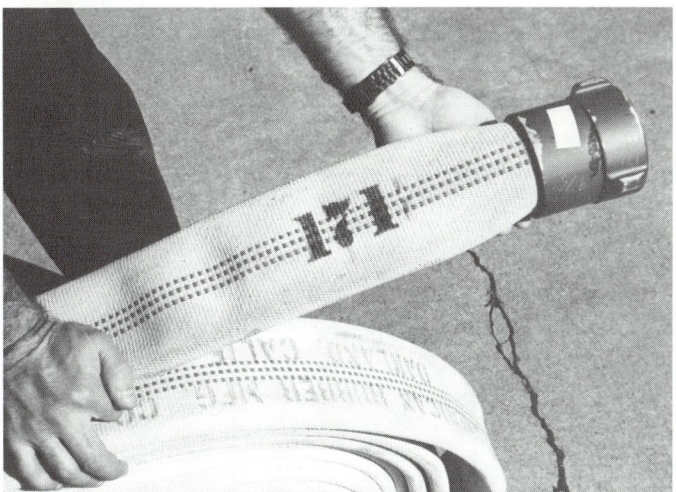
Figure 2.76 Stencil the identification number on the hose jacket with indelible ink.

A hose record is a case history of the section of hose from the time it is purchased until it is taken out of service. Records can be kept on cards, log sheets, or on computers (Figure 2.77). Information should include the date of purchase, the name of manufacturer, the date and results of periodic testing, remarks concerning the tests, repairs, unusual features, and causes of failure, if any.

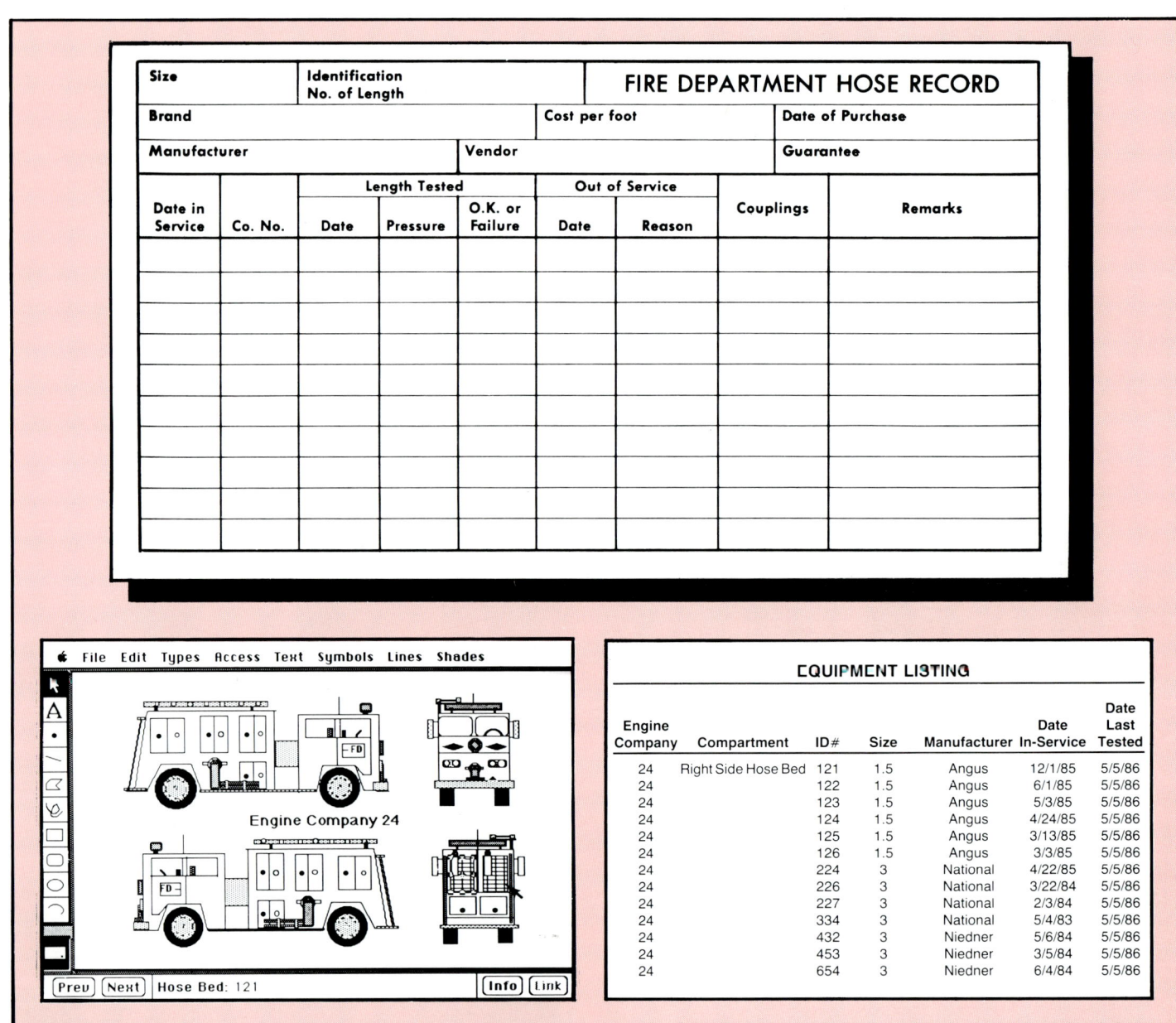

Figure 2.77 Sample hose records — on top, a hose card; on the bottom, a computerized record. *Courtesy of National Fire Hose Corp. and F.I.R.E. AID.*

Care, Maintenance, and Testing 57

Chapter 2 Review
Answers on page 233

MULTIPLE CHOICE: Circle the correct answer.

1. Damage to the inner lining of the hose would be most likely to result when the hose is _____.
   A. loaded on edge
   B. run over by a vehicle
   C. subjected to water hammer
   D. dragged behind a moving apparatus

2. If dirty hose is loaded on an apparatus, which type of damage would be most likely?
   A. Heat
   B. Chemical
   C. Mechanical
   D. Stress

3. The recommended method of cleaning a hose by hand is to _____.
   A. use a wire brush
   B. brush with a dry broom
   C. scrub with a broom and a water stream
   D. scrub with a hand brush and bleach solution

TRUE-FALSE: Mark each statement true or false. If false, explain why.

4. Precautions can be taken to prevent hose damage during training exercises, but there is no time for them during actual fire operations.
   ☐ T ☐ F _____

5. When a window must be broken to allow advancement of a hoseline, it is best to make as small a hole in the windowpane as possible.
   ☐ T ☐ F _____

6. Tools should not be carried on top of hose beds.
   ☐ T ☐ F _____

7. If hose must be dragged over pavement, it is best to drag it on edge to minimize abrasion damage.
   ☐ T ☐ F _____

8. If it is necessary for a vehicle to drive over a fire hose, it is better for the hose to be charged rather than uncharged.
   ☐ T ☐ F _____

58 HOSE

9. Hose should be removed from a hose tower as soon as it is dry.
   ☐ T ☐ F _____

10. Contact with hot objects can cause damage to the hose jacket but the liner will not be affected.
    ☐ T ☐ F _____

11. A musty smell is an indication of mildew on a hose, even if the mildew is not visible.
    ☐ T ☐ F _____

12. Fire hose can be damaged by chemicals in liquid form, but not by chemicals in gaseous form.
    ☐ T ☐ F _____

13. When possible, fire hose should be laid against curbs.
    ☐ T ☐ F _____

14. Most damage to couplings is due to water hammer or excessive pressure in the hose.
    ☐ T ☐ F _____

15. If it is necessary for a vehicle to drive over a fire hose, it is better to drive over a place well away from any couplings.
    ☐ T ☐ F _____

16. If it is necessary to drag hose, it should be folded in half and the couplings should be carried.
    ☐ T ☐ F _____

17. Once a coupling has been damaged in any way, it must be removed and replaced.
    ☐ T ☐ F _____

18. If a hose dryer or hose tower is not available, hose may be dried by laying the hose on edge in an apparatus room.
    ☐ T ☐ F _____

Care, Maintenance, and Testing

19. Hose in a drying tower should be allowed to swing in the wind to hasten drying.
    ☐ T ☐ F _____

20. Any time a fire hose is used, both the jacket and couplings should be inspected for damage before it is placed back in service.
    ☐ T ☐ F _____

21. During service testing of hose, test lengths of hose should not exceed 300 feet (90 m).
    ☐ T ☐ F _____

22. During service testing of hose, each test length of hose should be secured by a rope or hose strap to a nearby anchor.
    ☐ T ☐ F _____

23. During service testing of hose, the discharge nozzles should be held below the level of the pump discharge while the hose is being charged with water.
    ☐ T ☐ F _____

24. During service testing of hose, the required test pressure is 50 psi (345 kPa).
    ☐ T ☐ F _____

25. During service testing of hose, the test pressure should be maintained for three minutes.
    ☐ T ☐ F _____

26. During service testing of hose, a ⅛-inch (3.2 mm) movement of newly installed couplings is allowable during initial testing.
    ☐ T ☐ F _____

27. Because unlined linen hose leaks under pressure, it cannot be pressure tested.
    ☐ T ☐ F _____

28. Each hose section should be marked with an identification number.
    ☐ T ☐ F _____

60 HOSE

MATCHING: Write the correct letter in the space provided.

29. Match equipment associated with coupling equipment to coupling attachment methods.

    _____ Allen wrench   A. Expansion ring
    _____ Tool kit   B. Screw-in expander
    _____ Expanders   C. Collar
    _____ Vise and key   D. Tension ring

FILL IN THE BLANK: Fill in the blanks with the correct values.

30. During vacuum-testing of hard suction hose, no more than _____ inches (_____ mm) of vacuum should be lost within _____ minutes.

SHORT ANSWER: Answer each item briefly.

31. If a hose roller is not available, is there any way to prevent damage to a hose being pulled over a windowsill? If so, how?
    _____
    _____

32. If it is not possible to route a hoseline around debris, what are two other actions which could be taken to prevent damage to the hose?

    A. _____
    B. _____

33. What is the function of a hose bridge?
    _____
    _____

34. A hose connected to a pumper that vibrates excessively can be damaged by rubbing against the street surface. What hose appliance can be used to prevent this damage?
    _____
    _____

35. If hose in a hose bed is not used, how many times per year does NFPA 1962 recommend it be rotated?
    _____
    _____

36. What is water hammer?
    _____
    _____

37. What causes water hammer?
    _____
    _____

Care, Maintenance, and Testing **61**

38. How can water hammer be prevented?
   _____
   _____

39. During cold weather, how can water be kept from freezing in a hose?
   _____
   _____

40. What is the correct treatment for a hose on which mildew has formed?
   _____
   _____

41. What is the correct treatment for a hose that has come in contact with battery acid?
   _____
   _____

42. What is the recommended method of cleaning a dirt-encrusted coupling?
   _____
   _____

43. How often should fire hose be service tested?
   _____
   _____

YES-NO: Indicate whether the following methods are recommended to break frozen coupling connections by writing Yes or No.

44. _____ Place couplings near apparatus exhaust pipes

45. _____ Thawing the coupling with a propane torch

46. _____ Pouring hot water over towel-wrapped couplings

Indicate whether pressure testing of a hose is recommended after each of the following circumstances by writing Yes or No.

47. _____ Hose has been frozen

48. _____ Hose has been subjected to mildew

49. _____ Hose may have been damaged by chemicals

50. _____ Couplings have been replaced on the hose

LISTING

51. List the four major ways by which a lined fabric-jacket fire hose can be damaged.

A. _____

B. _____

C. _____

D. _____

52. List the two methods for removing a hose frozen in ice.

A. _____

B. _____

53. List three methods for avoiding mildew on cotton hose.

A. _____

B. _____

C. _____

54. List the three types of expanders.

A. _____

B. _____

C. _____

55. List the two reasons for vacuum testing hard suction hose.

A. _____

B. _____

DISCUSSION QUESTIONS

In your experience, what has been the most common reason for hose failure in your department? How could this have been prevented?

_____

_____

_____

Would you recommend any changes in the methods presently used by your department for hose recoupling? Describe and justify the changes.

_____

_____

_____

Would you recommend any changes in the methods presently used by your department for service testing hose? If so, why do you feel these changes would be improvements?

_____
_____
_____
_____

LEARNING ACTIVITY

Write a standard procedure for preventive maintenance of the hose used by your department.

3

Hose Appliances and Tools

66 HOSE

This chapter provides information that addresses performance objectives described in NFPA 1001, *Fire Fighter Professional Qualifications* (1987), particularly those referenced in the following sections:

Fire Fighter I

3-13 Fire Hose, Nozzles, and Appliances

3-13.2

Fire Fighter II

4-13 Fire Hose, Nozzles, and Appliances

4-13.1

4-13.5

Chapter 3
Hose Appliances and Tools

As stated earlier, hose is used to transport water from one place to another. To control water as it is moved from the source to the fire, certain accessories must be used, such as shutoff valves and nozzles. These accessories are collectively identified as "appliances" and include nozzles, valves, and fittings. Hose appliances, therefore, are devices other than couplings that are used with hose and through which water must pass.

A number of tools and devices are also available that make the job of hose handling much easier. Just as importantly, they help prevent injury to hose through mishandling.

NOZZLES

A nozzle is a device that directs water from the hose to the fire. It forms the water into a fire stream and controls the stream so that fire is extinguished in the most efficient manner (that is to say, using a minimum amount of water, with the least amount of water damage). A nozzle generally consists of two parts: a shutoff valve and a tip. The shutoff valve provides a means of not only opening and closing the nozzle (turning the water on and off), but in some cases is a means of controlling the amount of water that flows through the tip. The tip is the component that forms the stream. It is a precisely engineered device that directs the water to the area of application, much the same as a rifle barrel directs a bullet to its target. In some basic designs, the shutoff valve and the tip are distinctly separate; in other designs, they are combined as inherent parts of the nozzle.

Solid Stream Nozzles

A solid stream nozzle is the oldest type of nozzle in the fire service. An advantage of this nozzle is that its stream can penetrate a mass of burning material when a fire is deep seated. Another advantage is that it forms a stream of water capable of reaching a great distance. This is of great value during a defensive attack when the fire is exceptionally hot and cannot be closely approached. A disadvantage with a solid stream nozzle is that, if not carefully directed, the stream can cause significant property damage because of the force of the water.

The simplest type of solid stream nozzle consists of a shutoff valve and a smoothbore tip. The inside diameter of the tip corresponds to a specific flow in gallons per minute (gpm) (liters per minute [L/min]) at a set nozzle pressure. One of the variations of the nozzle has a shutoff valve with several sizes of tips "stacked" in descending order (Figure 3.1). There are similar master stream nozzles. This

Figure 3.1 One type of solid stream nozzle has several sizes of tips stacked in descending order. *Courtesy of Elkhart Brass Manufacturing Company.*

permits the nozzle operator to remove one or more tips to acquire the appropriate flow volume. Another advantage with this type of nozzle is that the hoseline can be easily extended. In this case, the nozzle is closed, all tips removed, and the desired length of hose is connected to the shutoff. When the extended hose is in place with another nozzle attached, the shutoff is opened.

The smoothbore tip can be used on both handline nozzles and master stream devices such as monitors and ladder pipes. Nozzle pressure is dependent on the type of device used. Typically, smoothbore tips on handline nozzles deliver an optimum stream at 50 psi (345 kPa). An optimum solid stream is one that remains intact, without fragmentation, for a specific distance. This distance varies with the size of the tip and with the nozzle pressure. Master stream tips operate best at 80 psi (552 kPa). (For more information about nozzles, see the appropriate section in IFSTA, **Fire Streams**.)

Fog Nozzles

A fog nozzle, as the name implies, produces a fire stream made of small droplets of water that leave the tip in a spray or "fog" pattern. This nozzle is designed to produce water droplets at the optimum size to be vaporized when introduced into a heated atmosphere such as found within a fire-engulfed structure. Most fog nozzles can be adjusted from a wide-angle stream down to a narrow pattern that resembles a solid stream (the narrow fog stream is actually hollow) (Figure 3.2).

As with solid stream nozzles, fog nozzles are used on both handlines and master stream devices. Most fog nozzles are designed to operate at a pressure of 100 psi (690 kPa). This means that at the specified pressure the nozzle will produce a stream at the rated volume with water droplets of the optimum size for vaporization. Older models of fog nozzles, however, produce a larger volume of water in a wide-angle fog stream than in the narrow-angle stream (hence the name "mystery nozzles"). Some newer models automatically adjust to variations in nozzle pressure to produce a stream that contains droplets of optimum size. At a nozzle pressure lower than 100 psi (690 kPa), the volume will decrease but the nozzle will maintain a stream of the same reach and quality as a stream produced at the higher pressure.

Exposure Nozzles

An exposure nozzle is designed to protect a building or object from the heat generated by a nearby burning building. It produces a fan-shaped stream, called a water curtain, approximately 35 feet (11 m) wide and two stories high (Figure 3.3). Contrary to popular belief, a water curtain does not prevent radiated heat from reaching an exposure. It cools the exposure in two ways: by pulling cool

Figure 3.2 Most fog nozzles can be adjusted from a wide-angle stream down to a narrow pattern that resembles a solid stream. *Courtesy of Elkhart Brass Manufacturing Company.*

Figure 3.3 A water curtain nozzle produces a fan-shaped stream. *Courtesy of Elkhart Brass Manufacturing Company.*

air into the space between the burning building and the exposure (thus forcing heated air out of the area), and by cooling the surface when the stream is applied directly on the exposure.

Applicator Nozzles

Applicator nozzles are designed to direct a water stream directly over the surface of a burning object. They were originally designed for the United States Navy and Coast Guard but many have found their way into public fire departments. One basic design has a fog tip that can be removed so that an applicator pipe (Figure 3.4) can be attached. This lightweight, curved pipe is typically used for extinguishing engine fires in vehicles. In this situation, the hood of the burning vehicle is raised only enough to insert the tip into the engine compartment, then the water stream is directed over the surface of the burning engine. The lowered hood protects the firefighters from exposure to flames that would ordinarily flare up if the hood were raised (thus allowing an inrush of air to the fire).

Figure 3.4 An applicator pipe. *Courtesy of Akron Brass Company.*

Another type of applicator nozzle is the piercing applicator. This nozzle has an angled, case-hardened steel tip. With a sledge or flat-head axe, it can be driven through a wall, roof, or ceiling (Figure 3.5). Small holes bored in the tip permit impinging jets of water to spray outward into the fire area. This nozzle is extremely effective for quick placement of water on a fast-traveling fire above a ceiling or within a wall.

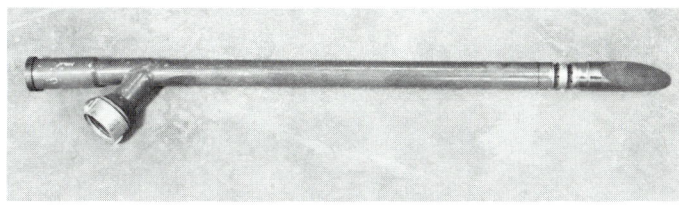

Figure 3.5 A piercing applicator nozzle.

Master Stream Devices

When fires grow to such a size and intensity that handline nozzles are incapable of providing enough water to gain control, larger devices must be used. Master stream devices (400 gpm [1 600 L/min] or more) include monitors, deluges, turret pipes, and ladder pipes. A monitor is a master stream device whose stream direction can be changed while water is being discharged (Figure 3.6). The stream straightener (also called a playpipe) is necessary because the water flow within

Figure 3.6 A monitor. *Courtesy of Elkhart Brass Manufacturing Company.*

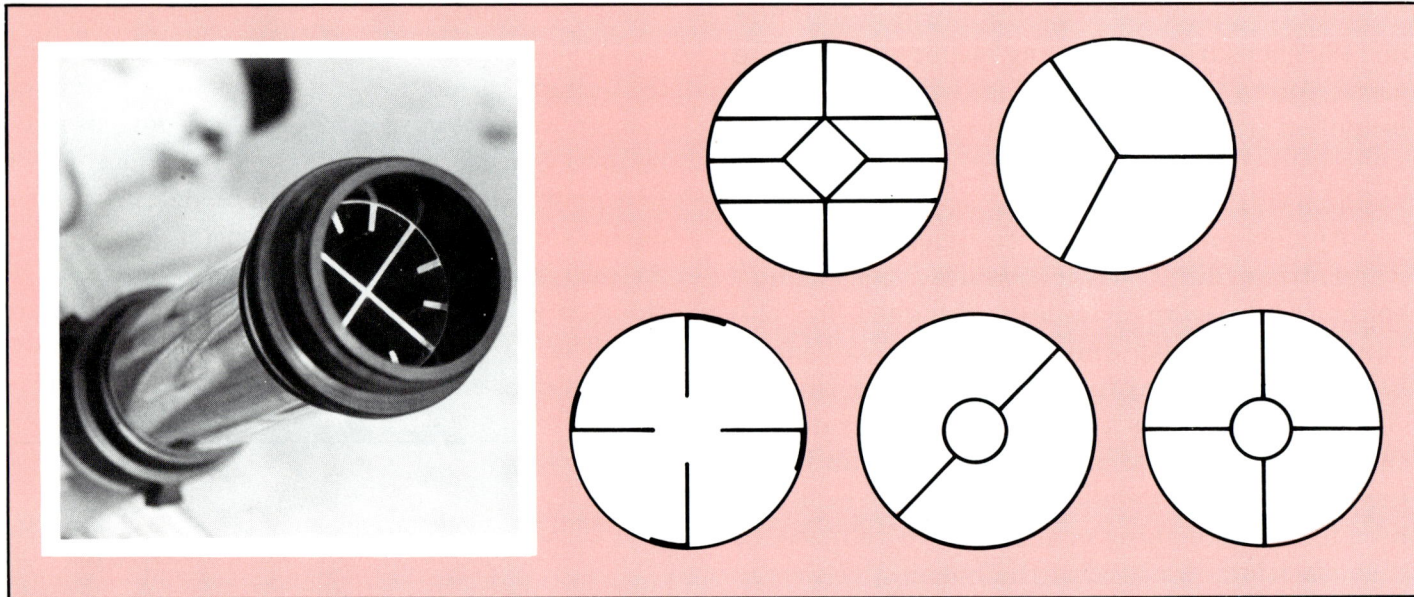

Figure 3.7 A playpipe (left) contains a vane pattern that smooths the flow of water. Several designs of vane patterns are used by playpipe manufacturers.

the appliance is turbulent as it approaches the nozzle. The straightener contains vanes that smooth the flow so that the nozzle discharges the water uniformly (Figure 3.7). The tripod-supported monitor, which has two or three inlets, is portable and may be operated from the top deck of a pumper or be placed on the ground. A turret pipe or deck gun is a large master stream appliance mounted on a pumper or trailer and connected directly to a pump (Figure 3.8). Such a device is capable of flows in excess of 2,500 gpm (9 464 L/min). A ladder pipe is a master stream device mounted on the fly of an aerial ladder (Figure 3.9). Some ladder pipes are portable, while others are preplumbed into the ladder superstructure.

Figure 3.8 A turret pipe (deck gun) is connected directly to the apparatus pump.

Figure 3.9 This ladder pipe is preplumbed to the aerial ladder. *Courtesy of Chico, California Fire Department.*

VALVES

When hose is connected to a fire hydrant or to a pump, there must be a means of controlling the water flow into the hose. This is accomplished with an appliance called a "valve." A valve is a device that contains an internal component that can be moved within the water passage to regulate the flow through the device. There are six basic types of valves, each named for this internal component:

- Gate
- Ball
- Butterfly
- Floating
- Clapper
- Piston

Gate Valves

The gate valve has a "gate" that moves into the water passageway when a crank handle is turned. One type of gate valve, known as a nonrising stem valve, has a double-walled gate that is threaded internally to allow a screw thread to move the gate into the waterway (Figure 3.10). Gate valves are relatively heavy and bulky but are well suited for use on a hydrant or pump discharge orifice.

Ball Valves

The ball valve is the most commonly used valve in the fire service. It is used in such equipment as pumper discharge orifices and gated intakes, in nozzles, and in "gated" wyes and siameses. As the name implies, the movable internal component of this valve is shaped like a ball (Figure 3.11). The ball has a hole through its center that permits water to flow through when aligned with the waterway. A quarter-turn of the handle rotates the ball to close the waterway, thus stopping the flow of water.

Figure 3.11 A ball valve.

Butterfly Valves

The butterfly valve has a disk that pivots within the waterway (Figure 3.12). It is open when

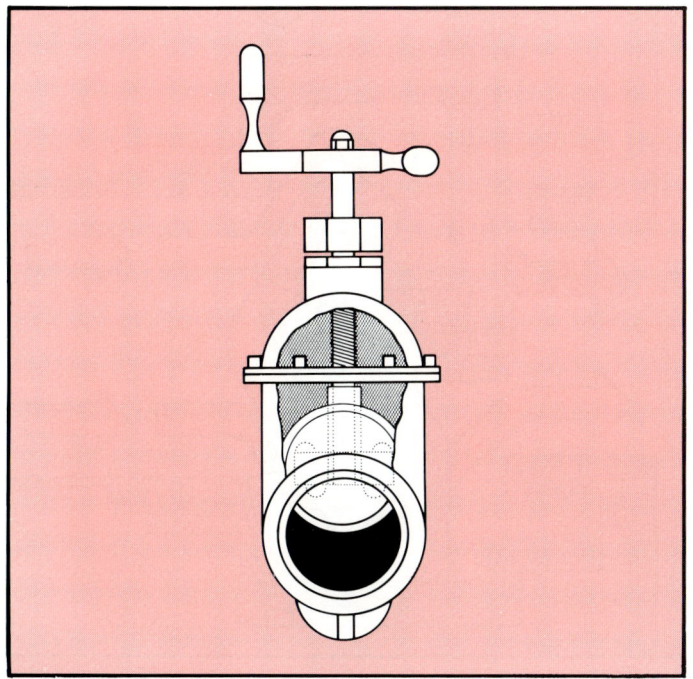

Figure 3.10 A gate valve.

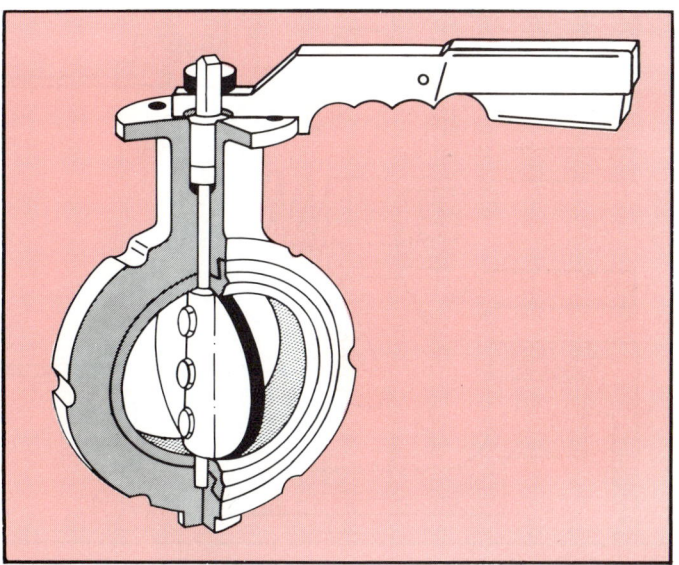

Figure 3.12 A butterfly valve.

the disk is aligned parallel to the waterway (the handle indicates this position by moving in the same plane). The valve is closed when the disk is turned a quarter-turn across the waterway. The butterfly valve, which is most frequently used on 4½-inch (115 mm) and larger pumper intake orifices, is not recommended for use on pump discharges. This is because, if not locked closed, the valve and its handle can be suddenly and violently thrown open from the force of the discharged water.

Floating Valves

A floating valve is a relatively new design. It is typically used on a main pump intake port to promote a quick intake hose connection. It has a spring-loaded, dome-shaped disk within the waterway that is held in the closed position by both spring tension and internal water pressure (Figure 3.13). When incoming pressurized water flows against the disk from the outside, it opens to permit water to flow through the valve.

Clapper Valves

A clapper valve is a hinged disk that acts as a check valve (Figure 3.14). It operates automatically by responding to waterflow direction within the waterway. A clapper valve opens when water flows in one direction and closes automatically when water flows in the opposite direction.

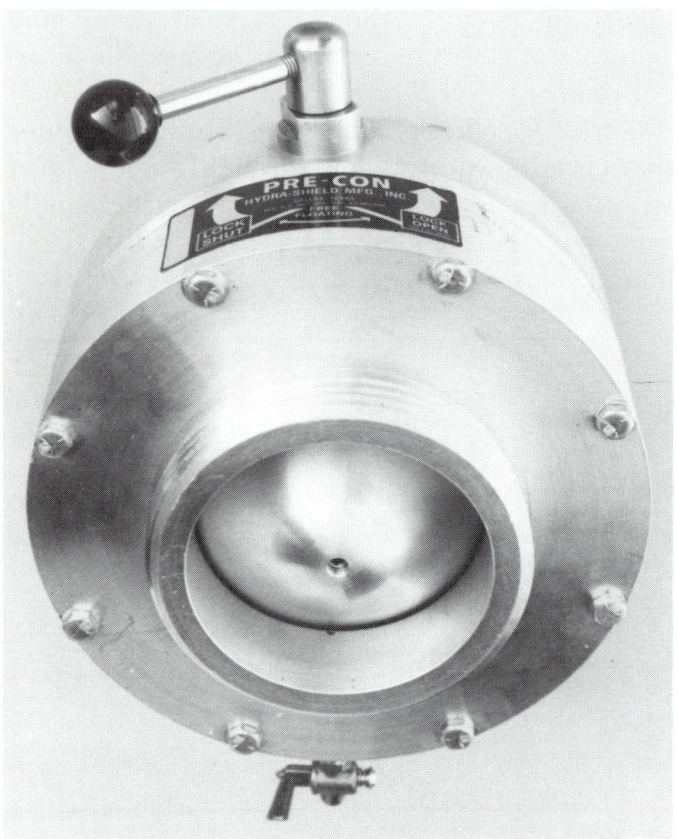

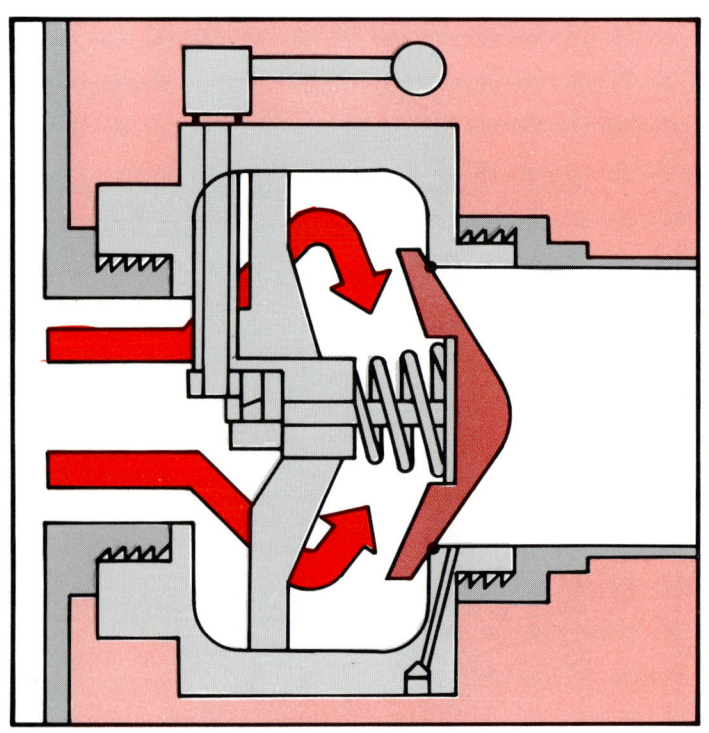

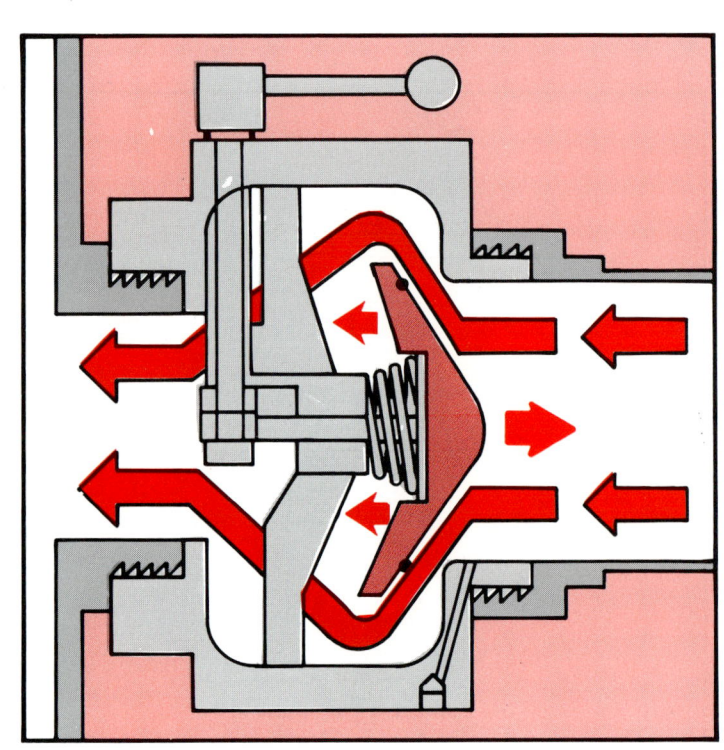

Figure 3.13 A floating valve. *Courtesy of Hydra-Shield Manufacturing, Inc.*

Hose Appliances and Tools

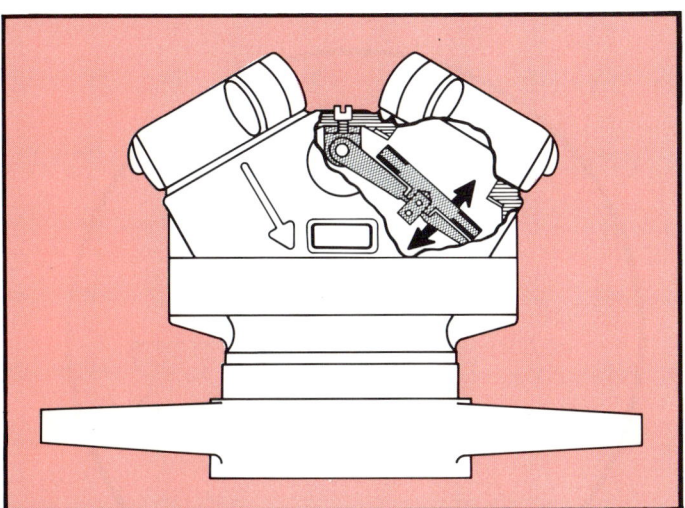

Figure 3.14 A clapper valve.

Piston Valves

A piston valve is designed for use with LDH hose. It has a piston that moves within a cylinder. The valve is opened and closed by rotating a wheeled handle. Rotating the wheel clockwise moves the piston downward into the waterway to block the opening. Counter-rotating the wheel retracts the piston from the waterway.

VALVE DEVICES

The movement of water between its source and the fire is often a complex operation, especially if the distance traveled is great. In many cases, the pressure in the hose must be increased somewhere along the hose lay to compensate for the pressure loss caused by water friction. In most cases, once a water flow is established it should remain uninterrupted by the pressure-boosting effort. A number of valve devices make this possible.

Four-Way Hydrant Valves

A four-way hydrant valve (also called a four-way gate or four-way valve) is a device that permits a pumper to increase the pressure to a supply hose between a hydrant and a pumper located some distance from the hydrant (Figure 3.15). More importantly, it provides the means to boost pressure without interrupting the water flow. This is especially important when pumping to fires that demand an increasing volume of water as the fire increases in size and intensity. The situation of connecting to a hydrant and later boosting the

Figure 3.15 Four-way hydrant valves (also called "four-way gates").

pressure is likely to occur when a single pumper arrives at a fire scene, lays its own supply hose from the hydrant to the fire, and remains at the fire. This is so the pumper can be utilized at the fire location as a source for tools and other equipment such as SCBA's and ladders. When a second pumper arrives, it is placed at the hydrant to boost

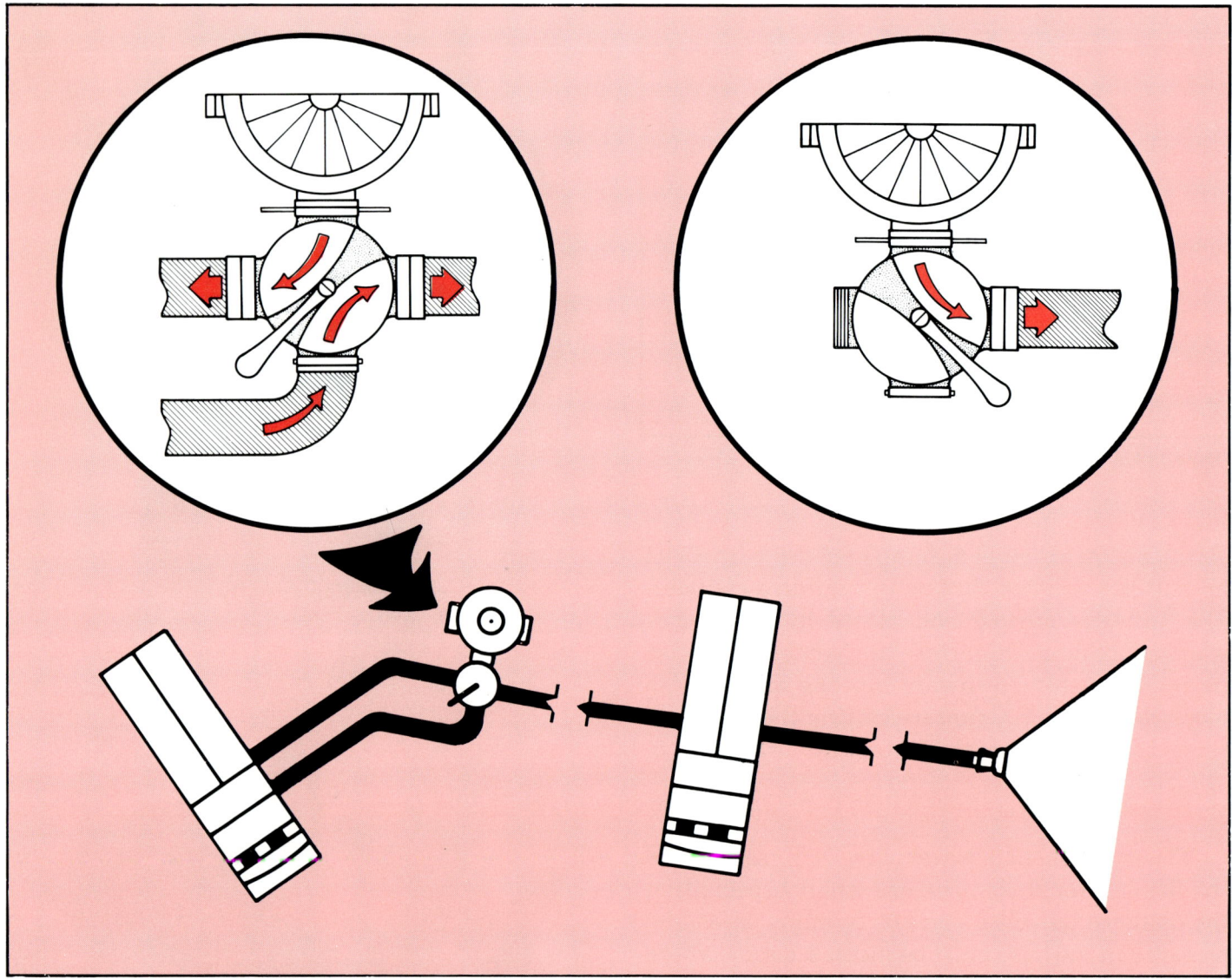

Figure 3.16 A four-way hydrant valve diverts hydrant water that is flowing directly into the supply hose (right inset). The valve reroutes the water to the pump, where it is pressurized, and returns it to the supply hose (left inset).

the supply hose pressure (it might also lay an additional supply hose if the situation demands). As shown in Figure 3.16, a simple four-way hydrant valve diverts hydrant water from the supply hose to the pump, where it is further pressurized. The water is then discharged back to the four-way valve, which directs it into the supply hose.

Automatic Hydrant Valves

When a pumper lays its own supply line from a hydrant to the fire location, someone must stay at the hydrant until the hose from the hose bed is disconnected and reconnected to an intake. When the connection is made, the hydrant is opened to charge the supply hose. This procedure, in effect, takes one person away from the fire scene for several minutes during the critical first minutes of initial attack on the fire.

An automatic hydrant valve eliminates the need to leave a person at the hydrant. The procedure for its use is the same as for any hydrant valve, except that the hydrant can be opened immediately, which permits the hydrant operator to rejoin the crew immediately. There are two kinds of automatic hydrant valves:

- Mechanically delayed
- Radio controlled

If the valve is the mechanically delayed type, it will automatically open after a preset period of time (Figure 3.17). This requires that *immediately* upon arrival at the fire scene someone either clamps the hose or makes an intake connection. If the hydrant valve is radio controlled, the pump operator transmits a radio signal through a special encoder to open the valve when the water is needed (Figure 3.18).

Manifolds and Water Thieves

In some situations, a pumper supplying several attack lines must be positioned some distance from the fire scene (perhaps pumping at a hydrant). There are two basic methods to provide water to each nozzle: pump directly to each attack line, or pump into a single hose that supplies a manifold. A manifold (or portable hydrant) is a device that receives a supply of water and distributes it through valves to a number of hoses (Figure 3.19). It is particularly useful in situations

Figure 3.19 A manifold distributes water to a number of hoses.

Figure 3.17 A mechanically delayed hydrant valve. *Courtesy of Hydra-Shield Manufacturing, Inc.*

Figure 3.18 A radio-controlled hydrant valve. *Courtesy of Frank T. Garza.*

where fire apparatus have limited access to buildings such as in shopping malls, large apartment complexes, and university campuses. Compared to laying a number of small diameter hoses, use of a manifold with LDH minimizes the amount of time required to lay hose and usually requires a lower pump pressure. An LDH-supplied manifold can support three or more 2½-inch (65 mm) attack lines. A disadvantage, however, is that if the supply line fails, all the attack lines connected to the manifold will be without water.

The water thief (Figure 3.20 on next page) is a variation of the gated wye. It has quarter-turn gate valves on the 1½-inch (38 mm) threaded outlets and may have a valve on the 2½-inch (65 mm) threaded outlet. The water thief is intended to be used on a 2½-inch (65 mm) or larger hoseline, usually near the nozzle. This allows hose with 1½- and 2½-inch (38 mm and 65 mm) threads to be used as needed for fire attack. The water thief can also be found with only 1½-inch (38 mm) threads.

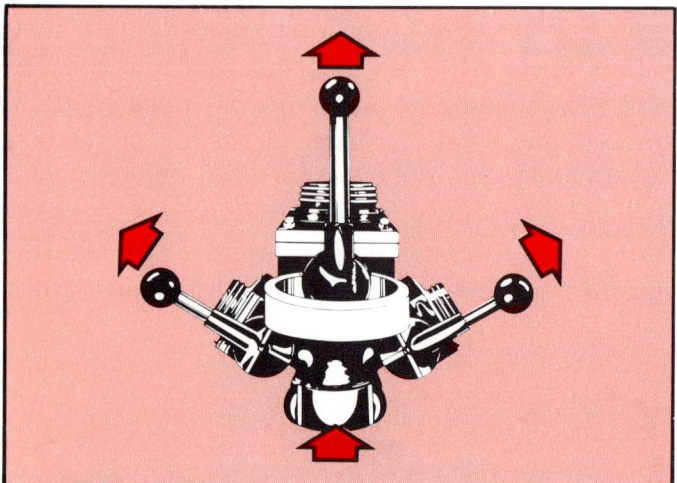

Figure 3.20 A water thief divides a single discharge line into as many as three lines.

Wyes and Siameses

When it becomes necessary to supply water directly to two attack hoses with a single line, one of the best appliances to use is a wye. A wye has a single female fitting on the input side and two male fittings on the output side. A gated wye has manually operated valves to permit separate control of water to each line. A reducing wye provides water for two attack lines smaller in diameter than the supply hose (Figure 3.21).

Figure 3.21 A reducing wye.

One method of reducing friction loss in a supply hose is to lay two parallel lines for a portion of the hose lay. The appliance for this purpose is a siamese, which has two female fittings on the input side and one male fitting on the output side (Figure 3.22). Most siamese appliances have check valves (clapper valves) so that if unequal pressure is transmitted through the two parallel lines, water from the hose of greater pressure will not return to the hose of lesser pressure. Friction loss in siamese fittings tends to be high; therefore, each fitting should be tested to ensure that accurate compensation can be made when pumping through them.

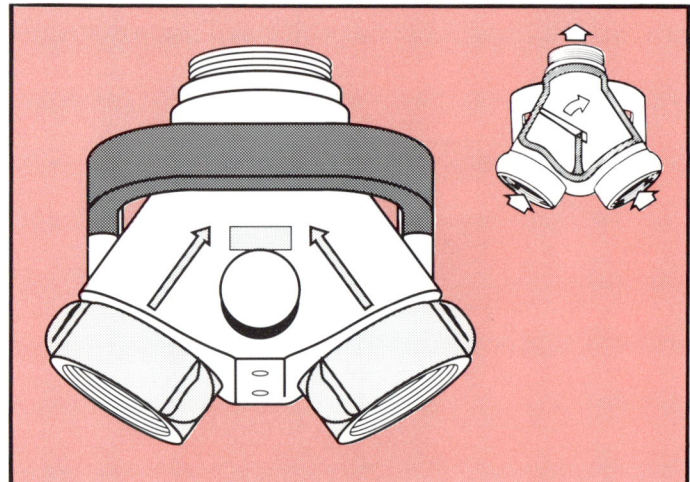

Figure 3.22 A siamese.

In-line Relay Valves

When a supply line is of exceptional length, friction loss can significantly deplete pressure, which reduces the amount of water delivered to the terminal end of the hose. One of the best ways to boost pressure in the hose is to position one or more pumpers along the line to "relay" the water. When a supply hose is laid prior to the arrival of the relay pumpers, in-line relay valves are placed at regular intervals within the lay (Figure 3.23). This allows the pumpers to connect to the supply hose and boost its pressure without interrupting the water flow.

Intake Relief Valves

The force of a water hammer in a large diameter hose is significantly greater than the force in smaller hose. For this reason, the potential for damage to a pump is also greater when LDH is used as a supply hose. One of the best ways to prevent this damage, of course, is to open and close valves slowly. Another way to prevent damage is to place a pressure relief device, such as an intake re-

Hose Appliances and Tools 77

FITTINGS

A number of more simple hardware accessories, called fittings, are available for connecting hoses of different sizes and thread types. When it becomes necessary to connect couplings of differing thread types, an adapter must be used. An adapter is a fitting for connecting hose couplings with dissimilar threads but with the same inside diameter. Adapters also connect threaded couplings to sexless couplings or to snap couplings. Double male and double female adapters are used to connect two threaded couplings of the same thread type, size, and sex (Figure 3.25). Reducers and increasers are adapters used to connect couplings of differing diameters (Figure 3.26). Elbows are used on pump

Figure 3.23 An in-line relay valve. *Courtesy of Jaffrey Fire Protection Company, Inc.*

lief valve, on the pump intake. An intake relief valve has a spring-loaded internal component that functions to divert any sudden pressure surge away from the pump (Figure 3.24).

Figure 3.25 A double male and double female adapter.

Figure 3.24 An intake relief valve. *Courtesy of Snaptite, Inc.*

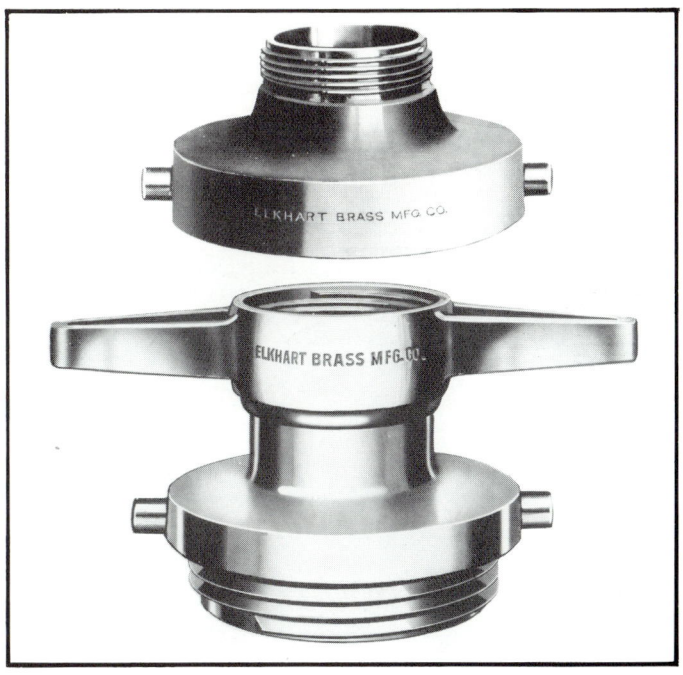

Figure 3.26 Increasers and reducers. *Courtesy of Elkhart Brass Manufacturing Company.*

intakes and discharge valves to reduce kinking of hoselines and to allow preconnection of suction sleeves (Figure 3.27). Caps and plugs are used to

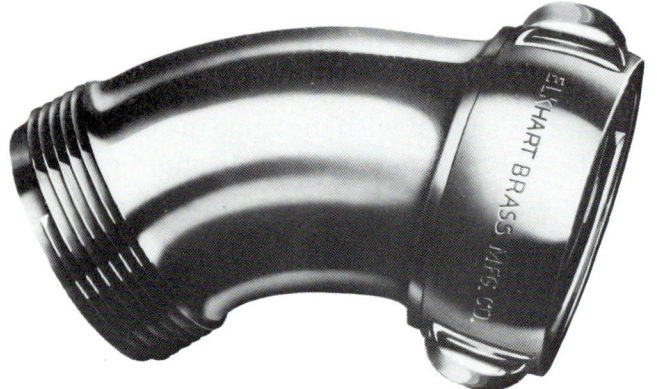

Figure 3.27 An elbow fitting. *Courtesy of Elkhart Brass Manufacturing Company.*

seal the ends of hoses and to protect the threads of discharge and inlet orifices on pumpers and other apparatus (Figure 3.28). Caps are for male fittings and plugs are for female fittings. Blindcaps seal the ends of hoses equipped with sexless couplings (Figure 3.29).

Figure 3.28 A threaded cap and plug. *Courtesy of Bar-Way Manufacturing Co., Inc.*

Figure 3.29 A blindcap seals the end of a hose equipped with a sexless coupling. *Courtesy of Snaptite, Inc.*

PROPORTIONERS AND EDUCTORS

An eductor is a portable proportioning device that injects a liquid, such as foam concentrate, into the water that flows through a hoseline. An in-line eductor (Figure 3.30) can be placed anywhere along a hoseline, but the length of hose between the eductor and the nozzle is limited by individual design. A foam nozzle eductor attaches directly to a special foam nozzle (Figure 3.31). All eductors work on the venturi principle (Figure 3.32). A low pressure area created by the rush of water past an internal opening draws the concentrate through a pickup tube from a portable container.

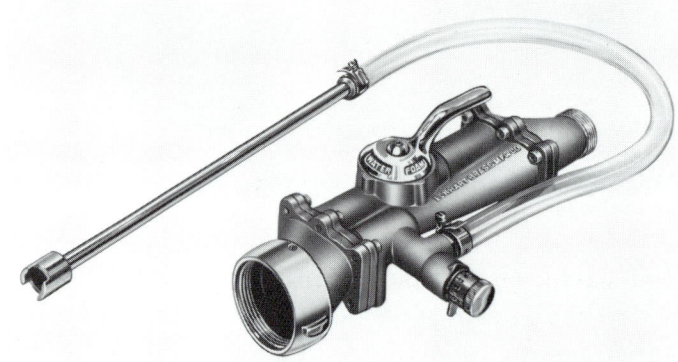

Figure 3.30 An in-line eductor. *Courtesy of Elkhart Brass Manufacturing Company.*

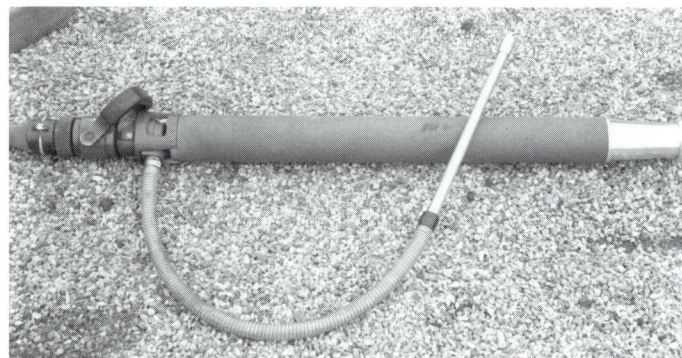

Figure 3.31 A foam nozzle eductor.

A proportioner is generally regarded as more of a fixed device than an eductor. A proportioner may be placed on a pumper discharge valve or may be permanently mounted behind the pump panel. With permanently mounted proportioners, the concentrate is stored in a separate storage tank plumbed directly to the proportioner and pump. Most modern proportioners and eductors have metering valves so that the percentage of liquid injected can be varied.

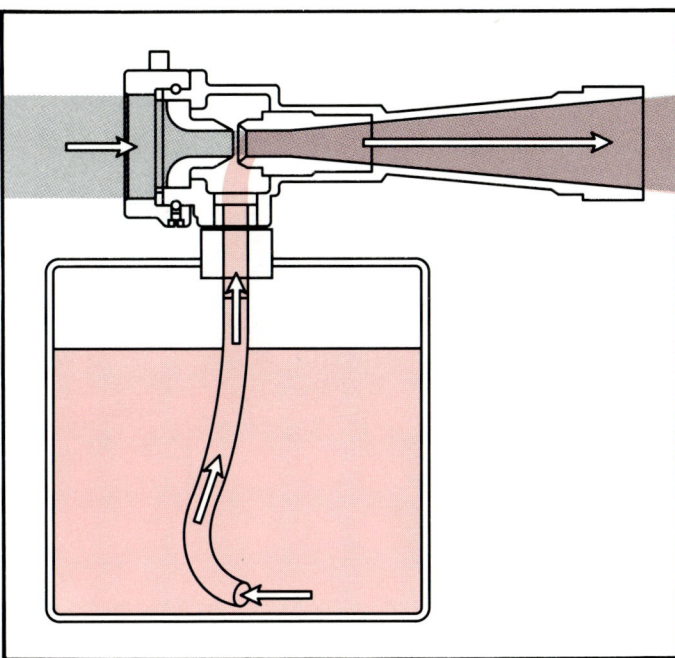

Figure 3.32 All eductors work on the venturi principle.

GENERAL CARE AND MAINTENANCE OF APPLIANCES

As with any fire fighting equipment, hose appliances should be well maintained to prevent failure under emergency conditions. The following maintenance rules apply to most appliances:

- Thoroughly inspect the appliance on a scheduled basis, as well as after each use.
- Never use grease or oil to lubricate moving parts. Use lubricants such as silicone or graphite that will not injure the hose jacket, hose lining, gaskets, O-rings, or other rubber or elastomer parts.
- At the onset of severe cold weather, relubricate all threads to prevent fittings from freezing together when coupled.
- Replace worn gaskets in female fittings. Carry spare gaskets on the apparatus.
- Make connections hand tight to avoid flattening the gaskets. If leaking connection must be tightened with a spanner wrench, tighten only enough to stop the leak.
- Carry appliances with valves open to prevent the accumulation of dirt and debris.
- Do not use brass polish or any other abrasive on hardened aluminum alloy.
- Repaint appliances as needed to prevent rust and corrosion.
- If an appliance is damaged and subsequently repaired, it should be service tested at 250 psi (1 724 kPa).

TOOLS AND OTHER DEVICES

There are a number of accessories available that make the handling of hose and appliances easier. Some devices help protect hose against unnecessary wear and damage.

Spanner and Hydrant Wrenches

The primary purpose of a spanner wrench, or spanner, is to tighten and loosen hose couplings (Figure 3.33). A number of other features have been built into some spanner wrenches:

- A wedge for prying
- An opening to fit gas valves
- A slot for pulling nails
- A flat surface for hammering

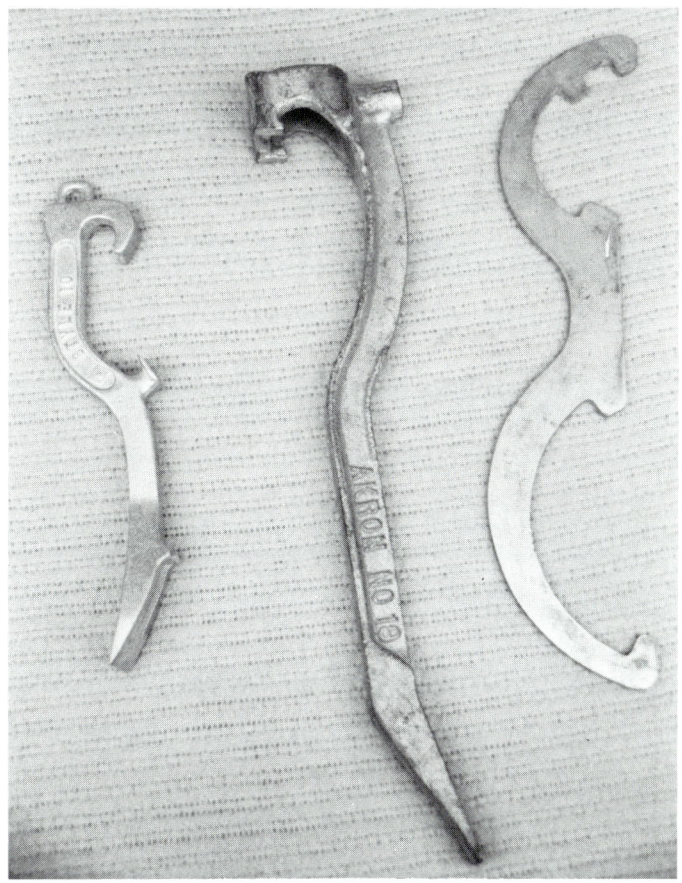

Figure 3.33 Spanner wrenches.

Hydrant wrenches are primarily used to remove caps from fire hydrant outlets and to open fire hydrant valves (Figure 3.34). Some hydrant wrenches are designed with an extension that can be used to tighten and loosen couplings.

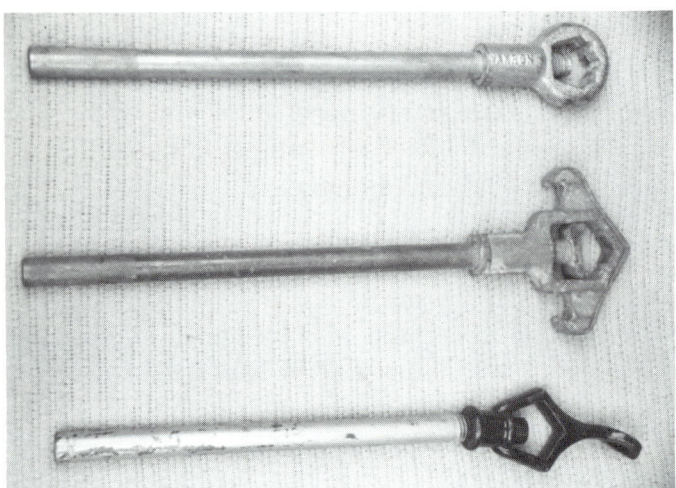

Figure 3.34 Hydrant wrenches.

Hose Straps, Ropes, and Chains

One of the most useful tools to aid in handling a charged hoseline is a hose strap. Similar to the hose strap are the hose rope and hose chain (Figure 3.35). These devices can be used to carry and pull fire hose, but their primary value is to provide a more secure means to handle pressurized hose when applying water. Another important use of these tools is to secure hose to ladders and other fixed objects.

Hose Control Device

In situations where a hoseline must be kept in a static position for an extended length of time, a hose control device holds the nozzle end of a 2½-inch (65 mm) hoseline (Figure 3.36). The device is stable enough to be left unattended as an exposure protection line when manpower is short or conditions are too hazardous for personnel to be in the area.

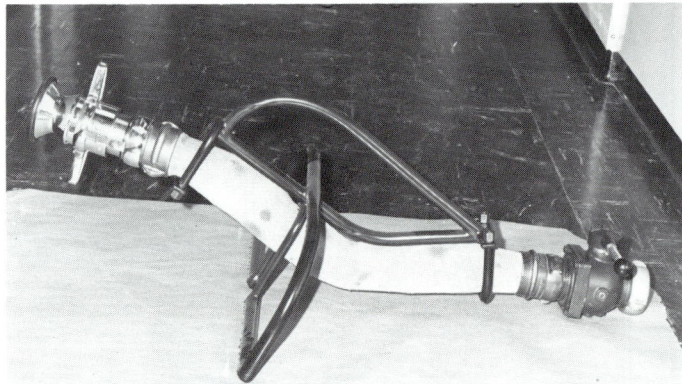

Figure 3.36 A hose control device can be left unattended to hold a working hoseline. *Courtesy of Louisville, Kentucky Fire Department.*

Hose Rollers

Hose can be damaged when dragged over sharp surfaces such as found on roof edges and windowsills. A tool for preventing such damage is the hose roller (Figure 3.37). Consisting of a metal

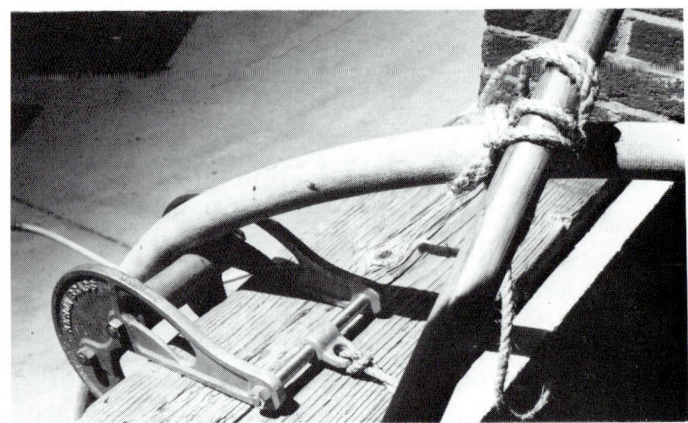

Figure 3.37 A hose roller prevents damage to hose from being dragged over rough or sharp edges.

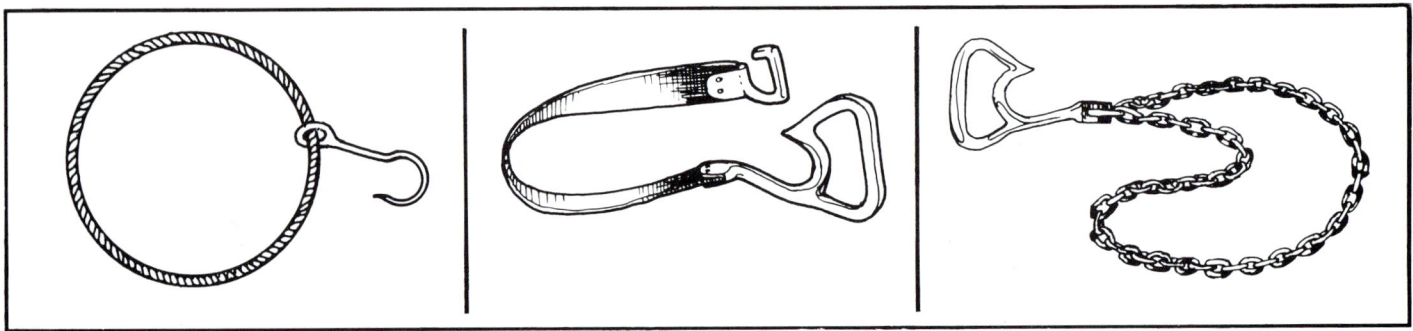

Figure 3.35 A hose rope, hose strap, and a hose chain.

frame with two or more rollers, the roller is placed on the potentially damaging edge and secured with a rope or C-clamp. The hose is then pulled over the hose roller.

Hose Jackets

When a section of hose ruptures, the entire hoseline is unable to transport water effectively. The most practical way to permanently correct the problem is to shut the line down and replace the damaged section of hose. When fire fighting conditions are such that it is not possible to shut down the hoseline and replace the bad section, a hose jacket can be installed on the hose at the point of rupture. The hose jacket encloses the hose so effectively that it can continue to operate at full pressure. A hose jacket can also be used to connect hose with mismatched or damaged screw-thread couplings.

A hose jacket consists of a two-piece metal cylinder that hinges open and closed (Figure 3.38). Rubber gaskets at each end of the cylinder seal against the hose to prevent leakage. A clamp device locks the cylinder closed when in use. Hose jackets are made in two sizes: 2½ inch and 3 inch (65 mm and 77 mm).

Hose Clamps

A hose clamp is used to stop the flow of water in a hoseline:

- To prevent charging the hose bed during hose layout operations
- To allow replacement of a burst section without shutting down the water supply
- To allow extension of a hoseline without shutting down the water supply

Based on the method by which they work, there are three types of hose clamps (Figure 3.39):

Figure 3.38 A hose jacket.

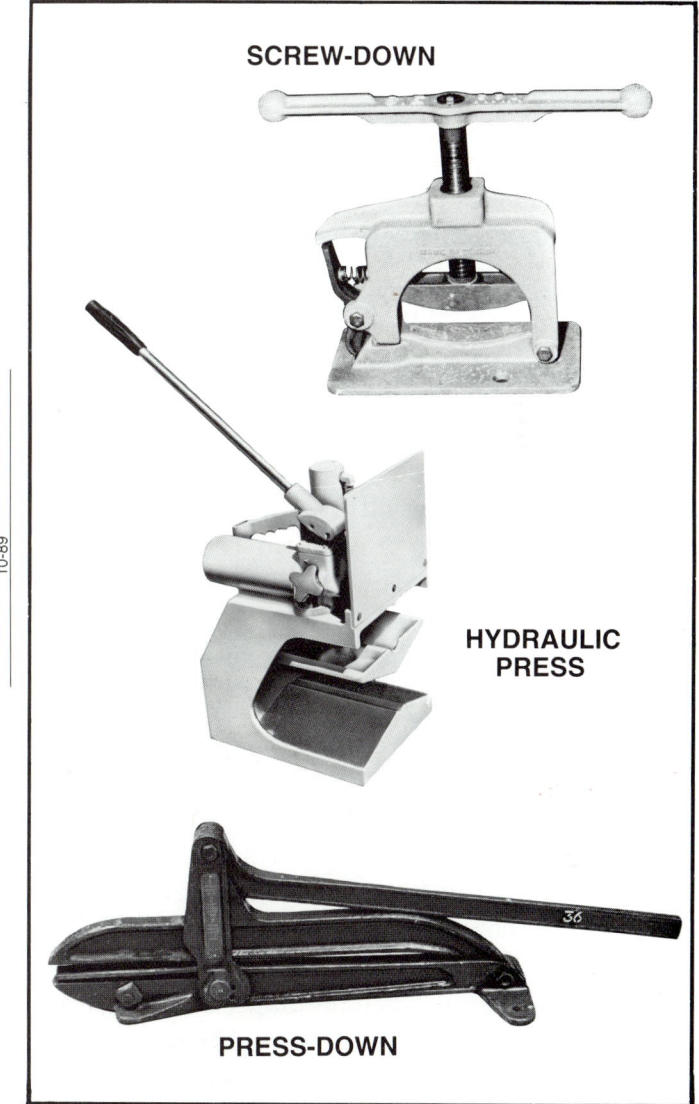

Figure 3.39 Three types of hose clamps: screw-down clamp, press-down clamp, and hydraulic press clamp.

82 HOSE

screw-down, press-down, and hydraulic press. It is important to remember that a hose clamp can cause injury to firefighters or damage hose if not used correctly. Some general rules that apply to all hose clamps are as follows:

- Apply the hose clamp at least 20 feet (6 m) behind the apparatus.
- Apply the hose clamp no farther than 5 feet (2 m) from the coupling on the incoming water side.
- Stand to one side when applying or releasing the press-down type of hose clamp (the operating handle is prone to suddenly snapping open).
- Place the hose evenly in the jaws to avoid pinching.
- Close and open the hose clamp slowly to avoid water hammer.

Suction Hose Strainers

Suction hose strainers are attached to the drafting end of a hard suction (sleeve) to keep debris from entering the fire pump. Such debris can pass through the pump and down the line to plug the nozzle. Figure 3.40 shows a floating strainer and a flat strainer.

Strainers should not be allowed to rest on the bottom of the water source except when the bottom is clean and hard, as with the bottom of a swimming pool. In order to prevent this, some strainers are provided with an eyelet to which a short length of rope can be attached. Some departments keep this rope attached to the strainer, as shown in Figure 3.41.

Hose Bridges

Hose bridges (also called hose ramps) help prevent injury to hose when vehicles cross it (Figure 3.42). They should be used whenever a hoseline crosses a street or other area where vehicular traffic cannot be diverted. Some nontraditional uses for hose ramps include placement over railroad tracks as crossing ramps and under hose as chafing blocks. Some ramps can also be positioned over small spills to keep hoselines out of potentially damaging liquids.

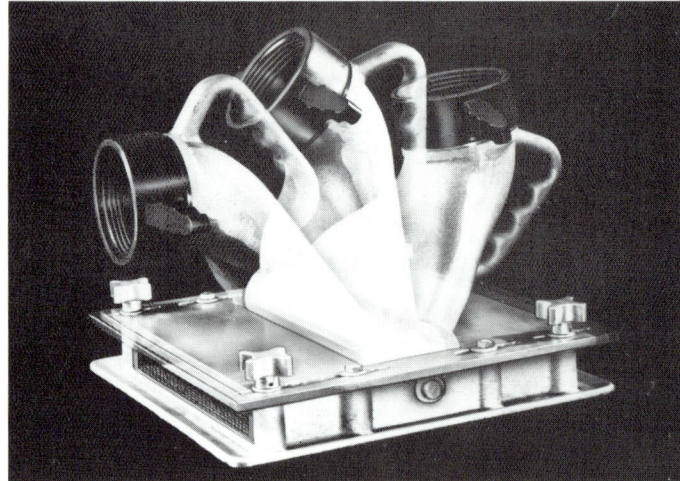
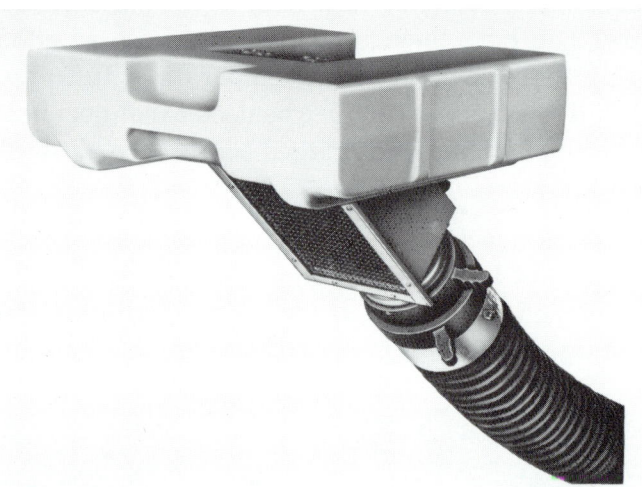

Figure 3.40 Two types of suction hose strainers: flat (top) and floating (bottom). *Courtesy of Ziamatic Corporation.*

Figure 3.41 A rope attached to a strainer helps keep the strainer from resting on the bottom of the water source.

Figure 3.42 Hose bridges help prevent damage to hose.

Hose Wringers

The hose wringer is used to remove water and air from LDH prior to reloading. Two firefighters place it over the hose and then walk down the hoseline, one on each side with the wringer clamped over the hose by one firefighter's grip (Figure 3.43).

Figure 3.43 A hose wringer removes water and air from large diameter hose prior to reloading.

Chapter 3 Review

Answers on page 234

MULTIPLE CHOICE: Circle the correct answer.

1. Which type of nozzle is especially well suited for extinguishing engine fires in vehicles?
   - A. Applicator
   - B. Exposure
   - C. Fog
   - D. Solid stream

2. Which fitting is used to connect couplings with dissimilar threads but with the same inside diameter?
   - A. Adapter
   - B. Blindcap
   - C. Elbow
   - D. Increaser

3. Which tool is primarily used to tighten and loosen hose couplings?
   - A. Hose clamp
   - B. Hose control device
   - C. Hydrant wrench
   - D. Spanner wrench

TRUE-FALSE: Mark each statement true or false. If false, explain why.

4. Hose appliances are defined as devices that are used with hose and through which water must pass.
   - ☐ T ☐ F _____

5. Couplings are classified as hose appliances.
   - ☐ T ☐ F _____

6. Solid stream nozzles are used on both handlines and master stream devices but fog nozzles are used only on handlines.
   - ☐ T ☐ F _____

7. Water curtains are effective because they prevent radiated heat from reaching an exposure.
   - ☐ T ☐ F _____

8. The piercing applicator nozzle is designed to be driven through a wall to allow water to be applied to a fire on the other side of the wall.
   - ☐ T ☐ F _____

9. The butterfly valve is well suited for use on pump discharge valves.
   - ☐ T ☐ F _____

10. The floating valve is well suited for use on pump intake orifices.
    ☐ T ☐ F _____
    _____

11. Four-way hydrant valves and in-line relay valves both provide a means of boosting pressure without interrupting water flow.
    ☐ T ☐ F _____
    _____

12. A gated wye has manually operated valves to permit separate control of water to each line.
    ☐ T ☐ F _____
    _____

13. It is not possible to connect two male couplings, even though they may be the same size and thread type.
    ☐ T ☐ F _____
    _____

14. Hose appliances should be carried on the apparatus with their valves open.
    ☐ T ☐ F _____
    _____

15. A hose clamp should be opened and closed slowly on a hose.
    ☐ T ☐ F _____
    _____

MATCHING: Write the correct letter in the space provided.

16. Match hose accessories to devices.
    _____ Eductor
    _____ Fitting
    _____ Manifold
    _____ Nozzle
    _____ Valve

    A. A device that directs water from the hose to the fire.
    B. A device that contains an internal component that can be moved to regulate the water flow through the device.
    C. A device that receives water and distributes it through valves to a number of hoses.
    D. A hardware accessory used to connect hoses of different sizes and thread types.
    E. A portable proportioning device that injects a liquid into the water that flows through a hoseline.

17. Match nozzle types to their descriptions.
    _____ Applicator
    _____ Exposure
    _____ Fog
    _____ Solid Stream

    A. Produces a stream that stays together as a mass.
    B. Produces a stream made of small droplets that leave the tip in a spray pattern.
    C. Produces a fan-shaped stream called a water curtain.
    D. Produces a stream to be applied directly over a burning surface.

86 HOSE

FILL IN THE BLANK: Fill in the blanks with the correct values.

18. Most smoothbore tips on handlines operate best at _____ psi (_____ kPa).

19. Most smoothbore tips on master stream appliances operate best at _____ psi (_____ kPa).

20. Most fog nozzles operate best at _____ psi (_____ kPa).

21. When applying a hose clamp to a hoseline, it should be applied at least _____ feet (_____ m) behind the apparatus and no further than _____ feet (_____ m) from the coupling on the incoming water side.

SHORT ANSWER: Answer each item briefly.

22. Which part of a nozzle stops and starts the flow of water?

23. Which part of a nozzle forms the stream?

24. Which type of master stream appliance is mounted on a pumper and connected directly to a pump?

25. Which type of master stream appliance is mounted on the fly of an aerial ladder?

26. Which type of valve is the most commonly used valve in the fire service?

27. Which type of valve opens when water flows in one direction and closes automatically when water flows in the opposite direction?

28. What is the function of a four-way hydrant valve?

Hose Appliances and Tools **87**

29. What is the function of an automatic hydrant valve?
   _____
   _____

30. What is the difference between a manifold and a water thief?
   _____
   _____

31. Which hose appliance is used to divide one line into two or more working lines?
   _____
   _____

32. Describe the fittings on the inlet and outlet sides of this appliance.
   _____
   _____

33. Which hose appliance is used to join two or more lines into one hoseline or device?
   _____
   _____

34. Describe the fittings on the inlet and outlet sides of this appliance?
   _____
   _____

35. What is the function of an intake relief valve?
   _____
   _____

36. What type of fitting is used to seal the end of a hose with a female coupling?
   _____
   _____

37. What type of fitting is used to seal the end of a hose with a male coupling?
   _____
   _____

38. What type of fitting is used to seal the end of a hose with a sexless coupling?
   _____
   _____

39. How can couplings of differing diameters be connected?
   _____
   _____

88 HOSE

40. What type of fitting is used on pump intakes and discharge valves to reduce kinking of hoselines and to allow preconnection of suction hoses?

41. What type of lubricant should be used on the moving parts of hose appliances?

42. What device can be used to hold the nozzle end of a hoseline that must be kept in a static position for an extended period of time?

43. What is the function of a hose jacket?

44. What is the primary function of a hose bridge or hose ramp?

LISTING

45. List the two sizes of hose jacket devices.
    A.
    B.

46. List the six basic types of valves used on fire hose appliances.
    A.
    B.
    C.
    D.
    E.
    F.

DISCUSSION QUESTIONS

If cost were not a factor, what master stream devices do you think your department should be equipped with? Why?

If cost were a factor, what master stream devices would you settle for? Why? What would you give up for the cost?

_____

_____

_____

_____

LEARNING ACTIVITY

Write a departmental procedure for preventive maintenance of the hose appliances used in your fire department.

4

Basic Methods of Handling Hose

92 HOSE

This chapter provides information that addresses performance objectives described in NFPA 1001, *Fire Fighter Professional Qualifications* (1987), particularly those referenced in the following sections.

Fire Fighter I

3-13 Fire Hose, Nozzles, and Appliances.

3-13.2

3-13.3

3-13.4

3-13.8

3-13.9

3-13.10

3-13.13

3-13.14

Fire Fighter II

4-13 Fire Hose, Nozzles, and Appliances.

4-13.1

4-13.2

Chapter 4
Basic Methods of Handling Hose

Connecting sections of fire hose is a simple task that requires a relatively small amount of training. Deploying hose at the scene of a fire, however, can be a complex task based on any number of variables: size of the fire, location of the water source, amount of water available, and personnel and equipment resources. No matter how complex a hose lay, however, the success of the operation depends upon executing a series of basic tasks: carrying or dragging hose, connecting couplings, attaching nozzles, installing appliances, and so forth. This chapter is devoted to the discussion and illustration of a number of widely accepted methods for handling hose. It is essential that these methods be learned before moving on to more complex operations.

MAKING AND BREAKING HOSE CONNECTIONS

The process of making a hose connection is, for the most part, a simple matter of screwing together a threaded male and female hose coupling or joining two sexless couplings to make a continuous water conduit. Under fire fighting conditions, however, hose must be connected quickly and efficiently. The need for speed and accuracy under adverse conditions requires training. This training should include practice not only in the specific techniques of connecting hose, but also in connecting hose to discharge gates, nozzles, master stream devices, sprinkler and standpipe systems, hydrants, and special appliances. Some basic rules apply when making most hose connections:

- On threaded hose, make all connections hand tight without the use of spanners (Figure 4.1) so that the hose can be disconnected later by hand. This is important be-

Figure 4.1 Make all connections hand tight so that the hose can be disconnected later by hand.

cause everyone at the fire scene does not always carry spanners. Spanner tight connections could create a delay when the person breaking the connection must stop to find some spanners to break the connection.

- When connecting any type of swiveled coupling, always check for the presence of a gasket. This can be quickly done with a visual check (Figure 4.2), or, if sight is obscured by darkness or smoke, by feeling for the gasket with a finger (Figure 4.3). Hand-tightened couplings that leak should be checked for worn gaskets rather than tightened with spanners.

- Connect sections of fire hose so that the hose edges are in the same plane (Figure 4.4). This practice makes hose easier to handle and easier to load.

Figure 4.2 Visually inspect the swivel to ensure the gasket is in place.

Figure 4.3 If darkness or smoke hampers vision, feel the inside of the swivel to ensure the gasket is in place.

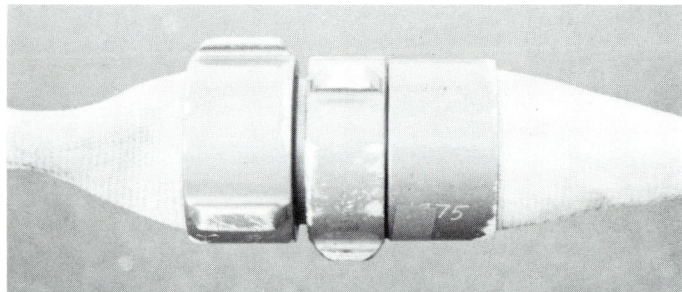

Figure 4.4 Fire hose should be connected so the edges are in the same plane.

Connecting Hose — One-Person Methods

Under fire fighting conditions it is common that a firefighter works alone to connect hose sections. Three methods are presented for connecting sections of hose without assistance. The same methods, in reverse order, can also be used to break the connections without assistance.

NOTE: Pictures of most hose handling methods are referenced to right-handed persons. In every case, left-handed persons can reverse the hand positions shown to make and break connections.

- *Coupling Tilt Method* — Stand at the side of the hose facing the two couplings. Place one foot on the hose directly behind the male coupling and apply pressure to tilt the coupling upward (Figure 4.5). A variant of this method is to start by making a short double fold of the male-coupled hose and then placing the coupling on the end of the fold (Figure 4.6). In either case, then pick up the female coupling of the opposite hose, place it against the upturned male coupling, and turn the swivel clockwise to make the connection (Figure 4.7).

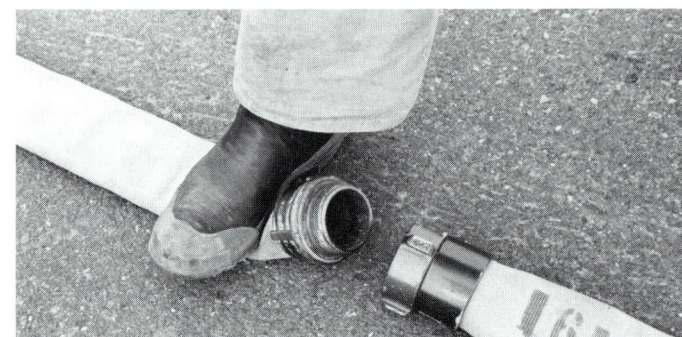

Figure 4.5 Step on the hose behind the coupling to tilt the coupling upward.

Basic Methods of Handling Hose **95**

Figure 4.6 Place the coupling on the end of a short double fold to aid in tilting it upward.

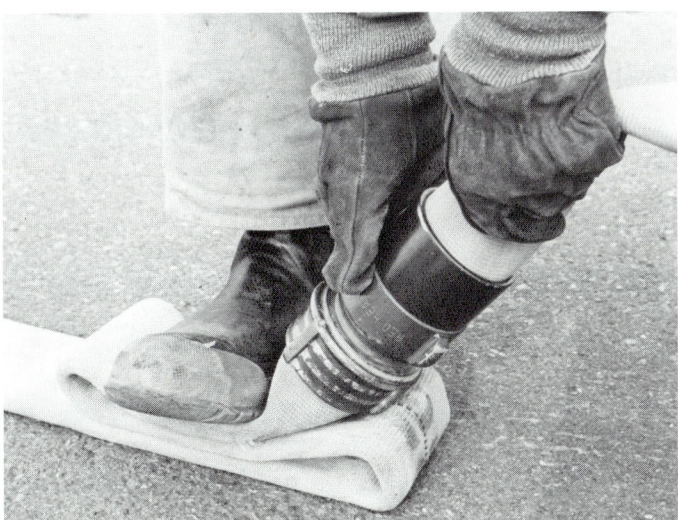

Figure 4.7 Place the female coupling against the upturned male coupling and turn the swivel clockwise to complete the connection.

- *Between-the-Feet Method* — Cradle the male coupling between the feet so that the coupling is tilted up. Pick up the female coupling of the opposite hose, place it against the upturned male coupling, and turn the swivel clockwise to make the connection (Figure 4.8).

- *Across-the-Leg Method* — This method is particularly suitable for connecting hose in bad weather because the couplings are lifted clear of snow, mud, or water on the ground. Grasp the female coupling in one hand with the hose straight behind, bend the knee on the corresponding side slightly, and lay the hose across the thigh. Cradle the male coupling in the opposite hand, bring the two couplings together, and turn the swivel clockwise to make the connection (Figure 4.9).

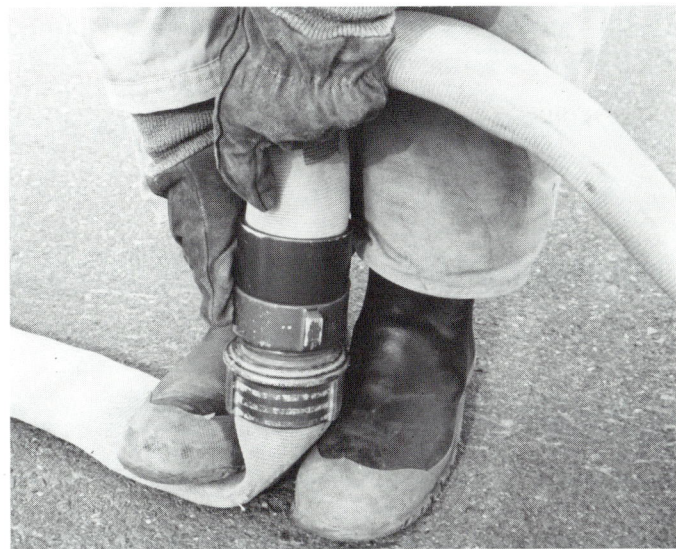

Figure 4.8 Cradling the male coupling between the feet, turn the swivel clockwise to complete the connection.

Figure 4.9 Lay the hose across one thigh to aid in connecting the couplings.

Connecting Hose — Two-Person Methods

When there are sufficient personnel available, two persons may work together to connect hose. The basic method starts with each person facing the other and holding, respectively, a male and female coupling. The person with the male coupling holds the coupling with the threads outward. A second person holds the female coupling shank with the swivel outward, then aligns the female coupling with the male coupling. (It is sometimes helpful for the person holding the male coupling to look away so that only one person aligns the couplings.) With the couplings aligned, the second person turns the swivel clockwise to make the connection (Figure 4.10).

Figure 4.11 Connect sexless couplings by aligning the couplings, inserting the lugs, and turning the swivels one-third of a turn clockwise to engage the lugs.

Figure 4.10 Align the male coupling with the female coupling and turn the swivel clockwise to complete the connection.

Connecting sexless couplings, such as Storz couplings commonly used with LDH, requires a slightly different method because both couplings have a swivel. In this instance, one person holds one of the swivels so that it will not turn. The other person aligns the couplings, inserts the lugs, and turns the swivel one-third of a turn clockwise to engage the lugs (Figure 4.11).

Breaking a Tight Screw-Thread Connection

It may sometimes become necessary to break a tight screw-thread connection when spanner wrenches are not immediately available. The methods presented are for both one and two persons.

- *Knee-Press Method (One Person)* — The principle of this method is that compression of the hose gasket permits the swivel to turn more easily. With the hose on the ground, grasp the hose behind the female coupling and stand the connection on end with the male coupling below. Best results will be obtained if the hose is bent close to the male coupling. Place one knee on the hose and shank of the female coupling, keeping the upper leg in a vertical plane with the couplings, and apply weight to the connection. At the same time weight is applied, quickly snap the swivel in a counterclockwise direction to break the tight connection (Figure 4.12).

- *Stiff Arm Method (Two Persons)* — This method uses the principle of leverage. In this case, arms held in a straight, rigid position act as levers to move the hands more effectively than if the arms were bent. Two persons start by facing each other with the hose coupling between them, their feet about shoulder width apart. Taking a firm, two-handed grip on the coupling, one person holds the male shank and the other person holds the female swivel. The fingers may be locked together to tighten the grip (Figure 4.13). Hold the arms rigidly straight. Starting with the body leaning slightly to the right, each person should apply force by leaning to the left while keeping the arms rigidly straight (Figure 4.14). This levering action produces a counterclockwise force on the couplings. The weight of the body should supply most of the force for breaking loose the connection.

Figure 4.14 Holding your arms rigidly straight, use the body to break the tight connection.

- *Spanner Wrench Method* — Spanner wrenches can be used to break tight connections when hand methods will not work. Place one spanner across the lugs on the swivel of the female coupling; place a second spanner across the lugs on the shank of the male coupling (Figure 4.15). Holding the male coupling spanner firmly in place,

Figure 4.12 Applying weight to the couplings, snap the swivel counterclockwise to break the tight connection.

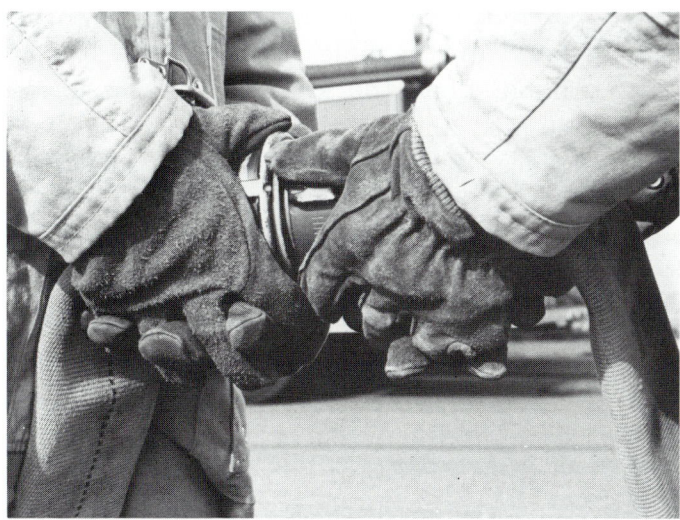

Figure 4.13 Interlock the fingers to tighten the grip.

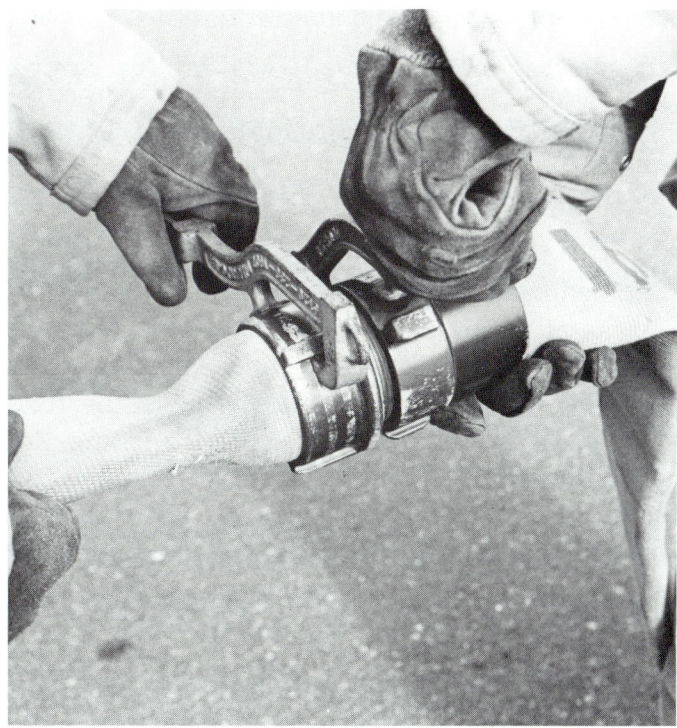

Figure 4.15 Place one spanner against the lugs of the female swivel and a second spanner against the lugs of the male coupling.

rotate the female coupling spanner in a counterclockwise motion to loosen the connection (Figure 4.16).

Figure 4.16 To break the connection, rotate the female coupling spanner counterclockwise while holding the male coupling spanner stationary.

Connecting Hose to Fixed Fittings

The procedure for connecting a female hose coupling to a *fixed* male fitting, such as a pump discharge valve, is similar to that used to connect a female coupling to a male hose coupling. In this case, simply place the female coupling against the male thread, aligning the coupling with the fitting, and turn the swivel clockwise to complete the connection.

Connecting a male hose coupling to a *fixed* female fitting, such as that found on the auxiliary intake at the pump panel, requires a slightly different procedure. In this instance, place the male coupling against the female swivel of the fixed fitting, aligning the coupling with the fitting, and turn the swivel *counterclockwise* to complete the connection (Figure 4.17). This procedure is also appropriate for fire department standpipe and sprinkler connections, which have female fittings. If these fittings have breakaway caps, fracture the caps by striking them sharply with a spanner or other striking tool. Clear away the broken cap pieces, then attach the hose with a counterclockwise rotation of the swivel. **NOTE:** Before attaching hose to a standpipe or sprinkler connections with the caps missing, look and feel inside for debris. If debris is present, remove it before making the hose connection.

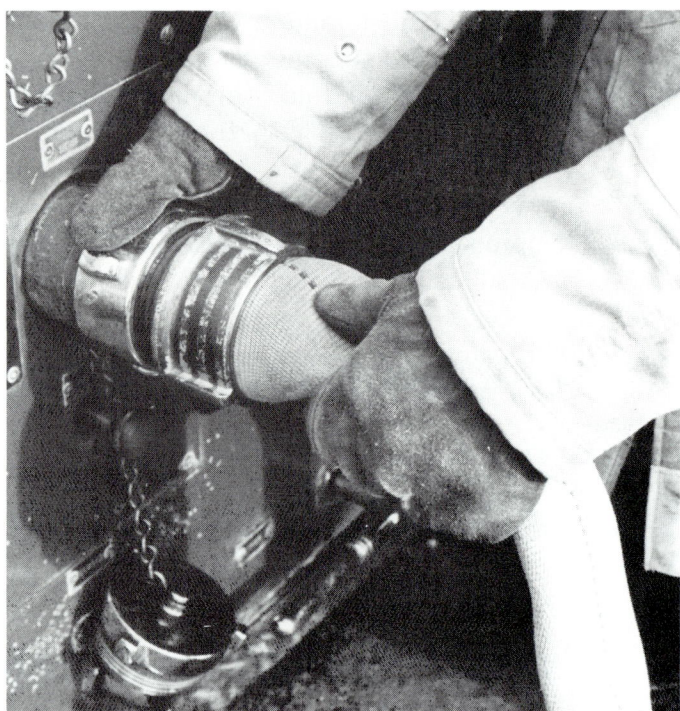

Figure 4.17 Place the male coupling against the female swivel of the auxiliary intake and rotate the swivel counterclockwise until the connection is complete.

Attaching a Nozzle to Hose

Nozzles may be attached to fire hose couplings by methods similar to those used for connecting couplings. Most nozzles have female threads and so attach to the male end of the hose.

- *Method One* — Hold the nozzle firmly in one hand and the male coupling in the opposite hand. Bend the knee on the nozzle side slightly and rest the coupling on the thigh. Place the nozzle threads in alignment against the coupling threads, then turn the nozzle clockwise to complete the connection (Figure 4.18).

- *Method Two* — Place one foot on the hose directly behind the male coupling and apply pressure to tilt the coupling upward. Holding the nozzle in both hands, place it against the upturned male coupling and turn the nozzle clockwise to make the connection (Figure 4.19).

Basic Methods of Handling Hose **99**

Figure 4.18 Rest the coupling on the thigh, align the coupling and nozzle threads, and turn the nozzle clockwise to complete the connection.

Figure 4.19 Tilt the coupling upward with one foot behind the male coupling. Align the nozzle on the coupling, then turn the nozzle clockwise to complete the connection.

HOSE ROLLS

There are a number of different methods for rolling hose. Each roll has advantages over other rolls, based on its intended use. In all methods, care must be taken to protect all couplings.

Straight Roll

The straight roll is a one-person operation. It is suitable for hose that is going to be handled in one of the following ways:

- Placed in storage, especially rack storage
- Returned to quarters for washing
- Loaded back on apparatus at the scene (rolling the hose prior to reloading purges the air so the hose will load more compactly)

When hose with screw-thread couplings is rolled into a straight roll for storage, start rolling the hose at the end with the male coupling so that the coupling is protected within the core of the roll. If the roll is made to facilitate loading directly into the hose bed, make the roll so that the appropriate coupling appears at the outside of the roll (this will depend on how the hose bed is configured). The procedure for making the straight roll is as follows:

Step 1: Lay the hose out straight and flat on a clean surface (Figure 4.20).

Step 2: Roll the coupling over onto the hose to start the roll. Form a coil that is open enough at the center to allow the fingers to be inserted (Figure 4.21).

Figure 4.20 STEP 1: Lay out the hose in a straight line.

Figure 4.21 STEP 2: Form a coil, allowing space in the center for fingers.

Step 3: Continue to roll the coupling over onto the hose, forming an even roll. As the roll increases in size, keep its edges aligned on the remaining hose to make a uniform roll (Figure 4.22).

Step 4: When the roll is completed, lay it on the ground and tamp any protruding coils down into the roll (Figure 4.23).

Figure 4.22 STEP 3: Keep the edges of the hose aligned during the roll.

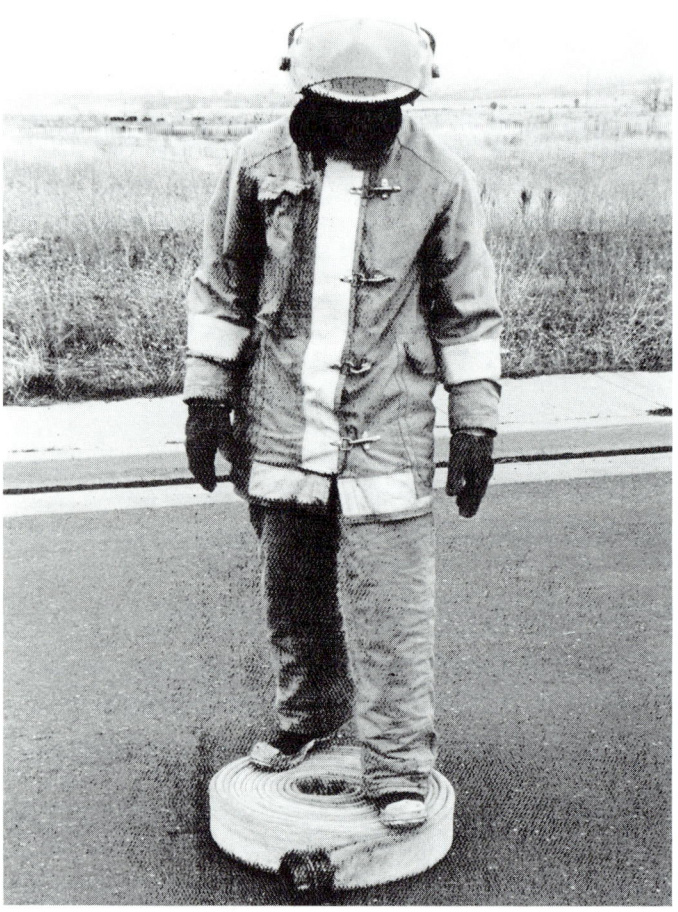

Figure 4.23 STEP 4: Tamp any protruding coils down into the roll.

NOTE: When rolling hose in a straight roll after a fire fighting operation, it is not usually necessary to drain the hose beforehand if it rests on a fairly level surface. The weight of the roll will compress the hose and push the water out of the line as the roll progresses. If the hose is rolled on a slope, roll the hose downhill whenever possible.

THE BUTTERFLY METHOD OF UNROLLING THE STRAIGHT ROLL

A traditional way of unrolling the straight roll is to push the roll away while grasping the outside coupling. The problem with this method is that the coupling at the opposite end can become damaged as it snaps downward against the ground from the momentum of the unrolling action. A method that avoids this damage is the butterfly method. It works especially well when space is limited. The procedure for unrolling the straight roll with the butterfly method is as follows:

Step 1: Unwind one layer of hose from the roll, then stand the hose roll on edge (Figure 4.24).

Figure 4.24 STEP 1: Unwind one layer of hose and stand the hose roll on edge.

Step 2: Grasp the first layer of hose on the roll, lift it enough to create some slack, and move it to one side of the roll (Figure 4.25).

Step 3: In the same manner, grasp the next layer (now the top layer) and move it to the opposite side. Continue pulling successive layers to alternate sides until the roll is deployed (Figure 4.26).

Step 4: Grasp the coupling from the center of the roll and walk away from the roll, laying the hose out on the ground (Figure 4.27).

Figure 4.25 STEP 2: Move the top layer of hose to one side.

Figure 4.26 STEP 3: Alternate the remaining layers from side to side.

Figure 4.27 STEP 4: Grasp the male coupling and walk away from the roll to lay out the hose.

Donut Roll

The donut roll is different from the straight roll in that both couplings are on the outside of the roll. This makes it easier to unroll the hose for loading into the hose bed or for extending a working hoseline. The couplings should be within 12 inches (305 mm) of each other in the completed roll. The hose can be deployed by holding both couplings and pitching the roll away so that it unrolls. Five variations of the donut roll are described here.

ONE-PERSON DONUT ROLL (METHOD ONE)

The first procedure for making the one-person donut roll is as follows:

Step 1: Lay out the section of hose in a straight line. Pick up the male coupling and carry it to the opposite end, placing it beside the female coupling. This will form a loop in the hose with two parallel segments (Figure 4.28).

Step 2: Go to the looped end and face the couplings. At about 30 inches (762 mm) from the loop on the male coupling segment, form a bight by lifting the hose enough to create a bend (Figure 4.29 on next page).

Step 3: Roll the bight over toward the male coupling to form a small coil, leaving enough space in the center to insert the fingers for carrying the finished roll (Figure 4.30 on next page).

NOTE: The 30-inch (762 mm) measurement is for 2½-inch (65 mm) hose; for 1½-inch (38 mm) hose, measure in about 18 inches (457 mm) from the looped end.

Figure 4.28 STEP 1: Lay out a section of hose in a straight line, then carry the male coupling to the opposite end to form two parallel segments.

102 HOSE

Step 4: Roll the hose so that it continues to form upon the male coupling segment. As the roll progresses, keep its edges aligned on the remaining hose to make a uniform roll (Figure 4.31). (If the hose in the female coupling segment becomes tight behind the roll, pull that hose back a short distance to create some slack.)

Step 5: When the roll approaches the couplings, lay it flat on the ground and draw the female coupling end around the male coupling end to complete the roll (Figure 4.32).

Figure 4.29 STEP 2: Lift the hose to form a bight.

Figure 4.31 STEP 4: Keep the edges of the hose aligned during the roll.

Figure 4.30 STEP 3: Roll the bight over to form a small coil. Leave room in the center of the coil to allow inserting the fingers of one hand.

Figure 4.32 STEP 5: Draw the female coupling end around the male coupling end to complete the roll.

Basic Methods of Handling Hose 103

ONE-PERSON DONUT ROLL (METHOD TWO)

A second procedure for making the one-person donut roll is as follows:

Step 1: Lay out the section of hose in a straight line. At a point 5 or 6 feet (1.5 m or 1.8 m) from the hose midpoint, lift the hose and form a bight. Roll the bight over toward the male coupling to form a small coil, leaving enough space in the center to insert the hand for carrying the finished roll (Figure 4.33).

Step 2: Continue to roll the hose so that the roll builds upon the male half of the outstretched hose. The female half of the hose will be dragged forward as it is pulled into the roll. As the roll progresses, keep its edges aligned on the outstretched hose to make a uniform roll (Figure 4.34).

Step 3: When the roll approaches the couplings, lay it flat on the ground and draw the female coupling end around the male coupling end to complete the roll.

TWO-PERSON DONUT ROLL

The procedure for making the donut roll with two persons is as follows:

Step 1: Lay out the section of hose in a straight line. Pick up the male coupling and carry it to the opposite end so that the hose is doubled back on itself. Place the male coupling on the hose 3 to 4 feet (0.9 m to 1.2 m) short of the female coupling (Figure 4.35).

Step 2: Straddle the hose at the looped end, facing the couplings. Pick up the looped end and

Figure 4.33 STEP 1: Lift the hose and form a bight, then roll the bight over toward the male coupling to form a small coil.

Figure 4.34 STEP 2: Keep the edges of the hose aligned during the roll.

Figure 4.35 STEP 1: Double the hose back on itself and place the male coupling 3 to 4 feet (0.9 m to 1.2 m) short of the female coupling.

roll the loop over to form a small coil, leaving enough space in the center to insert the hand for carrying the finished roll (Figure 4.36).

Step 3: Roll the hose upon the doubled hose, keeping its edges aligned on the hose to make a uniform roll. The second firefighter should pull the slack hose back as it appears ahead of the roll and maintain alignment of the doubled hose segments (Figure 4.37).

Step 4: Complete the roll. The female end should be longer than the male end so that it protects the male threads (Figure 4.38).

Figure 4.36 STEP 2: Roll the loop over to form a small coil.

Figure 4.37 STEP 3: Pull the slack out of the top layer of hose ahead of the roll. Maintain alignment of the doubled hose.

Figure 4.38 STEP 4: Complete the roll. The female end should be longer than the male end so that it protects the male threads.

TWIN DONUT ROLL

The twin donut roll is smaller in height but wider than a standard donut roll. Although best used for 1½-inch (38 mm) hose, larger sizes of attack hose can also be rolled in this manner. The procedure for making the twin donut roll is as follows:

Step 1: Lay out a section of hose so that the hose is doubled with the couplings together. Cross the two hose segments at a point approximately one-third along their length, then cross back the segments at a point two-thirds along their length (Figure 4.39). (The crossovers will bind the twin donuts together as they are rolled.)

Step 2: Roll the looped end over to form two small coils with both segments of hose (Figure 4.40). Leave enough space in the center to push through a carrying rope after the twin roll is completed.

Step 3: Roll the hose so that it continues to form a double roll upon both outstretched segments. As the roll progresses, keep its edges aligned on the outstretched hose to make a uniform twin roll (Figure 4.41).

Basic Methods of Handling Hose **105**

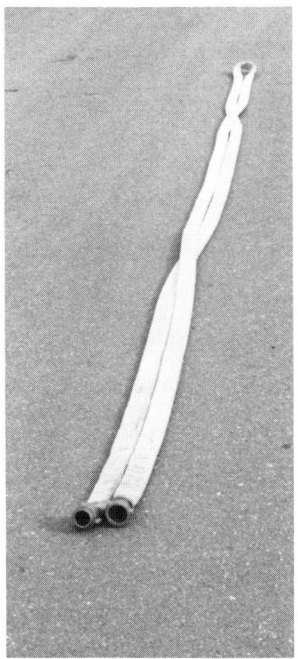

Figure 4.39 STEP 1: Cross the two hose segments at a point approximately one-third along their length, then cross back the segments at a point two-thirds along their length.

Figure 4.40 STEP 2: Roll the looped end over to form two small coils.

Step 4: Roll the completed roll over so that the couplings are on top. With the roll in an upright position, thread a strap through the center. Make a quick-release loop in the strap for carrying the roll (Figure 4.42).

Figure 4.41 STEP 3: Keep the edges of the hose aligned during the roll.

Figure 4.42 STEP 4: Secure the roll with a strap or rope tied with a quick-release loop.

106 HOSE

SELF-LOCKING TWIN DONUT ROLL

The self-locking twin donut roll is a twin donut roll that has a built-in carrying strap formed from the hose itself (Figure 4.43). This strap locks over the couplings to keep the roll intact for carrying. The length of the carrying strap may be adjusted to accommodate the height of the person carrying the hose.

Figure 4.43 The self-locking twin donut roll has a built-in carrying strap.

The procedure that follows is for making the self-locking twin donut roll:

Step 1: Lay out the section of hose in a straight line, then stand at the male end. Firmly grasp the male coupling (Figure 4.44) and rotate it around to the side so that it points toward the opposite end (Figure 4.45). This will create a bowl-like loop that is free of twists. *Do not* allow the coupling to twist in your hand.

Step 2: Maintaining a firm grasp on the coupling so that it does not twist, walk toward the opposite end on the same side of the hose as the male coupling. After walking 10 feet (3 m), step over to the opposite side of the hose to make a crossover and loop. Continue walking to the end and place the male coupling beside the female coupling (Figure 4.46).

Figure 4.44 STEP 1: Firmly grasp the male coupling.

Figure 4.45 STEP 1 (continued): Rotate the coupling around to the side so that it points toward the opposite end (do not allow the coupling to twist in your hand). This will form a bowl-like loop, free of twists.

Figure 4.46 STEP 2: Carry the male coupling to the opposite end. About 10 feet (3 m) along the hose, step over to the opposite side to form a crossover.

Step 3: Walk back to the looped end. As you walk back along the hose, move the hose segments together to lie uniformly in a straight line. At the looped end, adjust the crossover point of the two hose segments to be the proper distance from the loop (Figure 4.47) (this will be based on the anticipated size of the loop).

Step 4: Roll the loop over and place it on the crossover point (Figure 4.48).

Step 5: Holding the loop and crossed-over segments together, roll the hose forward to form two tight coils on the parallel segments (Figure 4.49).

Figure 4.47 STEP 3: Adjust the crossover point of the two hose segments to be the proper distance from the loop.

Figure 4.48 STEP 4: Roll the loop over and place it on the crossover point.

Figure 4.49 STEP 5: Holding the loop and crossed-over segments together, roll the hose forward to form two tight coils on the parallel segments.

Step 6: Roll the hose so that it continues to form a double roll upon both outstretched segments. As the roll progresses, keep its edges aligned on the outstretched hose to make a uniform twin roll (Figure 4.50).

Step 7: Complete the roll so that the couplings are on top. Pull the center loop to one side of

Figure 4.50 STEP 6: Keep the edges of the hose aligned as the roll progresses.

the roll so that the loop on that side becomes larger than that of the opposite side (Figure 4.51).

Step 8: Slip the long loop through the short loop on top of the roll, then push the loops down on the hose just behind the couplings (Figure 4.52).

Figure 4.51 STEP 7: Pull the center loop to one side of the roll so that the loop on that side becomes larger than that of the opposite side.

Figure 4.52 STEP 8: Insert the long loop through the short loop, then tighten the loops down against the top of the roll.

HOSE CARRIES AND DRAGS

One of the first steps for putting hose into service is to move it from the apparatus to the place of use. Smaller sizes of hose can be moved easily by one person, while larger sizes often require more than one person. Sections of hose are usually moved in two basic ways: carrying and dragging.

Hose Carries

Carrying is the preferred method because the hose is less subject to abrasion than when it is dragged. A general rule when making a preconstructed bundle for carrying is that the farther the bundle must be carried, the more it should be designed for comfort while carrying. In the methods shown, it is permissible to tie the hose bundles with rope or straps to make them more secure for carrying. Another general rule is that if speed of deployment is a factor, the bundle should be designed to lay out in a progressive manner with little chance of hanging up on obstacles or becoming entangled.

SHOULDER LOOP CARRY

The shoulder loop carry can be used by one person to carry a single section of hose or by several persons to carry a number of interconnected sections. The procedure shown here is for several persons:

Step 1: Lay out the desired amount of hose in a straight line. Pick up either end of the hose and place it over the shoulder with the coupling behind the body at waist height (Figure 4.53).

Step 2: Step forward to form a 3-foot (0.9 m) loop on the ground behind you and pick up the hose at the point closest to your feet (Figure 4.54).

Step 3: Raise the loop and lay it on your shoulder without twisting or turning it over (Figure 4.55). Expect that the resulting loop will have a half-twist in it. Adjust the loop so that the bottom of the loop hangs at about mid-calf.

Step 4: Step forward to form another 3-foot (0.9 m) loop on the ground behind you; as before, pick up the hose and lay it over the previous shoulder loop from behind. Continue to form successive loops in the same manner until the desired amount of hose is loaded (Figure 4.56).

Basic Methods of Handling Hose 109

Figure 4.53 STEP 1: Place one end of the hose over the shoulder so that the coupling hangs at waist height.

Figure 4.54 STEP 2: Step forward to form a 3-foot (0.9 m) loop on the ground behind you and pick up the hose at the point closest to your feet.

Figure 4.55 STEP 3: Raise the loop and lay it on your shoulder without twisting or turning it over. Expect that the resulting loop will have a half-twist in it. Adjust the loop so that the bottom of the loop hangs at mid-calf.

Figure 4.56 STEP 4: Continue to form successive loops until the desired amount of hose is loaded.

Step 5: Step forward approximately 15 feet (5 m) so that a second person may load shoulder loops in the same manner (Figure 4.57).

Step 6: Repeat the loading process with each person until the entire length of hose is loaded (Figure 4.58).

Figure 4.57 STEP 5: Move forward 15 feet (5 m) to allow the next firefighter to load shoulder loops in the same manner.

Figure 4.58 STEP 6: Repeat the loading process with each person until the entire length of hose is loaded.

Laying Out Hose Carried in Shoulder Loops

Laying out hose carried in shoulder loops by several persons is relatively simple but requires coordination and communication. As it is laid out the hose pays off the top of each person's shoulder to the rear.

To lay out hose carried in shoulder loops, the group walks to the place where the hose is to be connected to the water supply (for instance, at a pump discharge valve). Either connect the rearmost coupling in the load to the water supply or have someone hold the coupling to anchor the hose. The group then walks toward the destination point while maintaining a firm hold on their loops. The hose will pay off the rearmost person's shoulder first. To aid in dropping the loops, the rearmost person should lift the top loop from the shoulder and drop it when the hose pulls taut against its anchor (Figure 4.59). As the last fold pays out, the rearmost person signals the next person to start dropping loops. Repeat the dropping process in the same manner — top loop, then the next loop, and so on. Repeat this process until the entire length of hose is laid out.

Figure 4.59 Lift the top loop from the shoulder and drop it when the hose pulls taut against its anchor.

ACCORDION SHOULDER CARRY

The accordion shoulder carry is particularly useful for carrying a large volume of hose because several persons can carry interconnected bundles, then lay out the hose directly from the bundles. This carry can also be used by one person to transport a section of hose. Three methods for the accordion shoulder carry are described. The first method starts with hose removed from the hose bed and laying on the ground. The other two methods start with the hose still in the hose bed — one loaded with an accordion load and the other loaded with a flat load. (Both loads are described in greater detail in Chapter 5.) Each person performing an accordion shoulder carry usually carries one section of hose.

Method One: From Hose on the Ground

This method requires that the hose be laid out in a straight line so that it can be picked up while walking along the line. This method is also used to pick up hose after a fire and works well to drain water from hose as it is picked up.

Step 1: Lay out the section of hose in a straight line. Pick up either end of the hose and place it over the shoulder with the coupling behind the body at waist height (Figure 4.60).

Figure 4.60 STEP 1: Place the hose over the shoulder with the coupling behind your back at waist height.

Step 2: Hold the hose in front of the body and, while walking slowly forward, form a loop that hangs at knee height in front of the body (Figure 4.61).

Step 3: Walk on, guiding the hose back over the same shoulder to form a loop that hangs at knee height behind the body (Figure 4.62).

Step 4: Continue to walk slowly down the hoseline, forming alternating loops in front of and behind the body in the same manner as before, until the entire section is picked up (Figure 4.63).

Method Two: From Hose in an Accordion-Loaded Hose Bed

When a hose bed is loaded with hose in an accordion load, a number of folds can be moved directly from the hose bed to the shoulder in what amounts to a ready-made bundle. This procedure will not work, however, when the apparatus has an extremely long hose bed in which the folds exceed 10 feet (3 m) in length. Folds longer than 10 feet (3 m) will drag on the ground when carried by a person of average height.

Because the hose sections are connected in the hose bed, this method can be used by several persons to carry and lay out the hose directly from the shoulder. Installation of a nozzle is optional but is included here. The procedure is as follows:

Step 1: Attach a nozzle to the end coupling of the hose. Facing the hose bed, grasp the nozzle and the folds up to and including the fold that has the next coupling (Figure 4.64).

Step 2: Pull the folds halfway from the bed and, grasping them at the halfway point of their length, rotate them 90 degrees so that the fold with the nozzle is on the bottom. Turn away from the hose bed and place the folds on one shoulder (Figure 4.65).

Step 3: Hold the bundle tightly on the shoulder and step down from the tailboard. Walk

Figure 4.61 STEP 2: Form a loop in front that hangs at knee height.

Figure 4.62 STEP 3: Form a loop in back that hangs at knee height.

Figure 4.63 STEP 4: Guide the hose over the shoulder, alternating the loops in front of and behind the body until the entire section is loaded.

away to clear the folds from the apparatus and stop approximately 15 feet (5 m) from the tailboard (Figure 4.66).

Figure 4.64 STEP 1: Grasp the nozzle and several folds.

Figure 4.65 STEP 2: Place the folds on one shoulder with the nozzle fold on the bottom.

Figure 4.66 STEP 3: Step down from the tailboard and stop approximately 15 feet (5 m) from the apparatus.

Step 4: In the same manner as the first person, a second person steps up on the tailboard and faces the load. This person grasps the running length (which leads to the previously made bundle) and the folds up to and including the fold that has the next coupling. Rotate the hose 90 degrees so that the running length is on the bottom. Turn away from the hose bed and place the folds on one shoulder (Figure 4.67).

Step 5: Hold the bundle tightly on the shoulder and step down from the tailboard. Walk to a point 15 feet (5 m) away as the first firefighter moves ahead the same distance.

Figure 4.67 STEP 4: Place several folds on one shoulder with the running length on the bottom.

114 HOSE

Figure 4.68 STEP 6: Step down from the tailboard and move away 15 feet (5 m) as the first firefighter moves ahead the same distance. Additional persons load the hose in the same manner.

Step 6: Additional persons — load the hose in the same manner until the required amount of hose is removed from the bed (Figure 4.68). Disconnect the hose from the hose bed.

Method Three: From Hose in Other Hose Bed Loads

A different technique is required when hose is taken from a load other than an accordion load such as a flat load, a horseshoe load, or even from a reel. In this case, one person stands on the tailboard to pull hose from the hose bed and helps load hose onto the shoulder of each person carrying the load. Instructions, therefore, are for the person standing on the tailboard. As before, installation of a nozzle is optional but is included here. The procedure, shown with a flat load, is as follows:

Step 1: Attach the nozzle to the end of the hose. Position a carrier at the tailboard facing in the direction of travel. Place the initial fold of hose over the carrier's shoulder so that the nozzle can be held at chest height. Bring the hose from behind back over the shoulder so that the rear fold ends at the back of the knee (Figure 4.69).

Step 2: Make a fold in front that ends at knee height and bring the hose back over the shoulder. Continue to make knee-high folds until an appropriate amount of hose is loaded. Have the carrier hold the hose to prevent it from slipping off the shoulder (Figure 4.70).

Figure 4.69 STEP 1: Load the first layer with the nozzle in front at chest height and the rear fold at knee height.

Figure 4.70 STEP 2: Continue to make knee-high folds until an appropriate amount of hose is loaded.

Step 3: Have the carrier move forward approximately 15 feet (5 m) and position another person at the tailboard. Load hose onto the shoulder in the same manner as before, making knee-high folds until an appropriate amount of hose is loaded (Figure 4.71).

Step 4: Repeat the loading process with each carrier until the desired length of hose is loaded. Then uncouple the hose from the hose bed and hand the coupling to the last carrier (Figure 4.72).

Figure 4.71 STEP 3: Load the second carrier in the same manner as the first.

Figure 4.72 STEP 4: Uncouple the hose from the hose bed and hand the coupling to the last carrier.

Laying Out the Accordion Shoulder Load

To lay out hose carried in an accordion shoulder load, the group walks to the place where the hose is to be connected to the water supply (for instance, at a pump discharge valve). Either connect the rearmost coupling in the load to the water supply or have someone hold the coupling to anchor the hose. The group then walks toward the destination point while maintaining a firm hold on the hose bundles. The hose will pay off the rearmost person's bundle as the top fold is pulled loose by resistance from the anchor point (Figure 4.73). As the last fold pays out, the rearmost person signals the next person to release the hose so that it can pay out. The top fold, then the next fold, and so on will be pulled loose from the next person's bundle as the group walks on. Repeat this process until the entire length of hose is laid out.

Figure 4.73 Walk toward the destination point so that the hose pays off the rearmost person's bundle.

MODIFIED ACCORDION SHOULDER CARRY

The accordion shoulder carry can be modified so that the couplings are protected inside the folds. This simple preconstructed bundle is also designed for moving 2½- or 3-inch (65 mm or 77 mm) hose long distances, as when moving hose up the stairs of a high-rise building. This bundle can be pre-folded, tied, and stored ready to use. The procedure for the modified accordion shoulder carry is as follows:

Step 1: Lay out the section of hose in a straight line. Grasp either of the two couplings and fold the hose back upon itself approximately 5½ feet (2 m), laying the coupling on top of the outstretched line (Figure 4.74 on next page).

NOTE: This distance should be approximately one-ninth of the section length because the finished bundle will have nine folds of equal length.

116 HOSE

Figure 4.74 STEP 1: Lay the coupling back on top of the outstretched line to make a 5½-foot (2 m) fold.

Step 2: Pick up the coupling on the opposite end and carry it to the folded end while forming a wide loop in the center section of the hose. Place the coupling approximately 4 feet (1 m) to the side of and even with the folded end. Make sure the hose is free of twists (Figure 4.75).

Step 3: Return to the opposite end and pick up the hose at the center of the wide loop. Without twisting the hose, bring it to the center and lay this fold even with the coupling (Figure 4.76).

Step 4: Return to the opposite end and pick up the two loops. Without twisting the hose, bring the loops forward until they are even with the coupling (Figure 4.77).

Step 5: Move the coupling end around so that this fold lies inside the adjacent fold (Figure 4.78). At the opposite end of the hose, lay the folded end coupling to the inside of its adjacent fold. Adjust the bundle so that each fold rests on edge and is the same length.

Figure 4.76 STEP 3: Bring the center of the loop forward and lay the fold down even with the coupling.

Figure 4.77 STEP 4: Bring the two loops forward until they are even with the coupling.

Figure 4.75 STEP 2: Bring the other end of the hose to the folded end, forming a wide loop, and place the coupling approximately 4 feet (1 m) to the side of and even with the folded end.

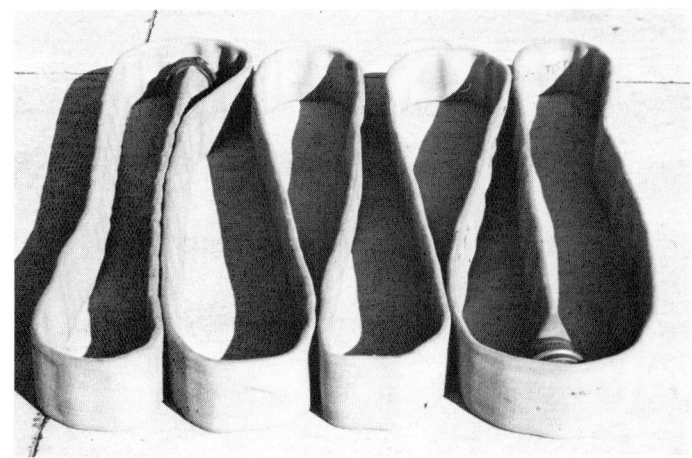

Figure 4.78 STEP 5: Move the coupling end around to lay inside the adjacent fold.

Step 6: Bind the completed bundle at each end with rubber bands, straps, or ropes tied with quick-release knots for storage and easy carrying (Figure 4.79).

Figure 4.80 STEP 1: Place the male coupling 12 inches (305 mm) to the side and even with the female coupling.

Step 2: Go to the opposite end and pick up the hose in the center of the loop. Carry the loop back between the couplings, and lay it on the ground even with the couplings (Figure 4.81).

Step 3: Go to the center of the folded hose and, placing one foot against the hose as a

Figure 4.79 STEP 6: Bind the completed bundle for storage and easy carrying.

HORSESHOE SHOULDER CARRY

The horseshoe bundle is useful for carrying a single section of hose on the shoulder or under the arm. The procedure for a horseshoe shoulder carry is as follows:

Step 1: Lay out the section of hose in a straight line. Pick up the male coupling and carry it to the opposite end, placing it approximately 12 inches (305 mm) to the side and even with the female coupling (Figure 4.80).

Figure 4.81 STEP 2: Place the loop of the hose between and even with the couplings.

pivot, swing the folded ends around and even with the couplings (Figure 4.82).

Step 4: Pick up the bundle at its center and place it on the shoulder (Figure 4.83).

HORSESHOE UNDERARM CARRY

A horseshoe shoulder bundle can be modified for carrying under the arm by adding one more step — folding the bundle once again to make a more compact load. The procedure for a horseshoe underarm carry is as follows:

Steps 1-3: Follow Steps 1, 2, and 3 of the procedure for the horseshoe shoulder carry.

Step 4: Go back to the center of the folded hose and, placing one foot against the hose as a pivot, swing the looped end around to the couplings so that the couplings are in the center of the bundle (Figure 4.84).

Step 5: Squat beside the completed bundle, compress the folds together, and stand the bundle on edge. Pick the bundle up and cradle it under one arm (Figure 4.85).

Figure 4.82 STEP 3: Placing one foot against the hose as a pivot, swing the folded ends around even with the couplings.

Figure 4.83 STEP 4: Pick up the bundle at its center and place it on the shoulder.

Figure 4.84 STEP 4: Swing the looped end around the couplings so the couplings are in the center of the bundle.

Basic Methods of Handling Hose **119**

Figure 4.85 STEP 5: Cradle the bundle under one arm.

Hose Drags

Dragging is a fast way to move hose from one place to another. While carrying is the preferred method for transporting hose, fire fighting operations sometimes require the firefighter to move hose from the apparatus to the fire scene with the greatest speed possible. For this reason, dragging is still a necessary method of handling hose.

The most simple method of dragging hose is to grasp both couplings and walk forward, dragging the doubled hose behind. If the hose is of a larger, heavier size, drape the hose ends over the shoulder with the couplings in front so that your upper body, rather than your arms, carries most of the hose's weight (Figure 4.86). *In no case should the couplings be dragged by pulling the hose from the folded end* (Figure 4.87).

An important fact to remember when dragging hose is that the less hose that contacts the ground, the easier it is to drag the hose. Most of the following methods take this into account and are designed to ease the dragging task.

Figure 4.86 When dragging a section of hose, carry the weight of the hose with the upper body instead of the arms.

Figure 4.87 NEVER drag the couplings by pulling hose from the folded end.

SINGLE SECTION (METHOD ONE)

This method is frequently known as the street drag. It starts with a section of hose laid out straight. The procedure is as follows:

Step 1: Standing to the side of one end of the hose, pick up the coupling with the hand

Figure 4.88 STEP 1: Walk the coupling to the midpoint of the hose section.

Figure 4.89 STEP 2: Pick up the hose at the midpoint and loop it over the shoulder.

nearest the hose, then walk to the midpoint of the hose section (Figure 4.88).

Step 2: With the other hand, pick up the hose at the midpoint and loop it over the shoulder nearest the hose (Figure 4.89).

Step 3: Step to the other side of the hose, crossing the chest with the hose, and walk to the opposite end (Figure 4.90).

Step 4: With the free hand pick up the second coupling. Walk forward to drag the hose (Figure 4.91).

Figure 4.90 STEP 3: Step to the other side of the hose, crossing the chest with the hose, and walk to the opposite end.

Figure 4.91 STEP 4: Pick up the second coupling and drag the hose.

Basic Methods of Handling Hose 121

SINGLE SECTION (METHOD TWO)

Another drag starts with a length of hose laid out straight. This procedure is especially well suited for dragging several interconnected sections of hose. Draping the body with hose as described distributes its weight equally over the upper body. The procedure, which is shown here with a shorter length of hose than usual, is as follows:

Step 1: Pick up a coupling and place it over the shoulder so that the hose end drapes across the chest (Figure 4.92).

Step 2: Walk to the other coupling, pick it up, and place it over the opposite shoulder so that the end crosses the chest in the opposite direction (Figure 4.93).

Step 3: Turn and step inside the loop that has been formed and walk back to the end of the loop. Pick up the loop and place it over the back of the neck so that the trailing hose drapes under the arms (Figure 4.94). Walk forward to your destination.

Figure 4.93 STEP 2: Lay the opposite end of the hose over the other shoulder so that it crosses the first end.

Figure 4.92 STEP 1: Pick up the coupling and place the hose over the shoulder with the end across the chest.

Figure 4.94 STEP 3: Pick up the loop and place it over the back of the neck so that the trailing hose drapes under the arms.

MULTIPLE SECTIONS (METHOD ONE)

When several interconnected lengths of hose must be moved by several people, one of the fastest ways is to pull the hose directly from the hose bed and immediately proceed to the area where the hose is needed. Each person drags one section of hose by grasping the connected couplings. To speed the process, one person stands on the tailboard to pull the hose from the hose bed. The procedure is as follows:

Step 1: If a nozzle is required, the tailboard person attaches the nozzle, then hands it to the first person, who stands facing away from the tailboard. This person lays the hose over the shoulder and holds the nozzle in front of the body (Figure 4.95).

Step 2: The nozzle carrier walks forward while the tailboard person pulls off hose. When a section of hose has been pulled off, the tailboard person calls for the nozzle carrier to stop and another person steps in on the same side of the hose as the previous person. The hose is placed over the shoulder with the coupling held in front, chest high (Figure 4.96).

Figure 4.96 STEP 2: The next firefighter shoulders the hose with the coupling in front of the chest.

Step 3: The tailboard person gives a verbal signal and the two carriers walk forward, dragging the hoseline, as the tailboard person pulls more hose from the hose bed. Each person pulls hose in the same manner until the required amount of hose is removed from the hose bed. When the coupling of the last section appears, the tailboard person stops the group, uncouples the last section from the hose bed, and either connects the hose to a discharge valve or carries it to the destination as the last carrier in the group (Figure 4.97).

Figure 4.95 STEP 1: Lay the hose over the firefighter's shoulder so the nozzle is in front of the body.

Figure 4.97 STEP 3: The last coupling can be connected to a discharge valve or carried to another location with the group.

MULTIPLE SECTIONS (METHOD TWO)

Another way for several persons to drag a number of interconnected sections of hose is to first pull the hose from the hose bed and lay it on the ground in folds. Each person then picks up several folds and proceeds to the destination. The procedure is as follows:

Step 1: If a nozzle is required, attach the nozzle to the end of the hose. Pull enough hose from the hose bed to allow placing the nozzle on the ground approximately 25 feet (8 m) behind the tailboard and on the side toward where the hose will be carried (Figure 4.98).

Step 2: Pull more hose from the bed and make a fold that starts at the tailboard and ends even with the nozzle (Figure 4.99). Lay this first fold beside the nozzle fold on the side away from where the hose will be carried.

Step 3: Continue to pull hose from the hose bed and make more folds in the same manner, laying each one progressively away from the nozzle in an accordion fashion. (Figure 4.100).

Step 4: When the appropriate amount of hose has been pulled, uncouple the hose from the hose bed. If the hose is to be connected to the pumper from which it was removed, do so at this time (Figure 4.101). If the hose is needed at another location, lay the last coupling along the last fold.

Step 5: The first person in the group stands at the nozzle facing away from the folds. Pick up

Figure 4.98 STEP 1: Pull the nozzle end from the hose bed and lay the nozzle about 25 feet (8 m) behind the tailboard.

Figure 4.100 STEP 3: Continue to lay folds, accordionlike, progressively away from the nozzle.

Figure 4.99 STEP 2: Pull more hose and make a fold that starts at the tailboard and ends even with the nozzle.

Figure 4.101 STEP 4: Break the hose connection at the hose bed and connect the hose to the discharge gate. (Pumper is relocated for clarity.)

Figure 4.102 STEP 5: Pick up the nozzle in one hand and the first fold in the other hand.

the nozzle, lay the hose over the shoulder, and hold the nozzle with one hand in front of the body at about chest height. Pick up a fold in the opposite hand and walk forward approximately 25 feet (8 m) (Figure 4.102).

Step 6: A second person moves in to stand at the end, facing away from the next two folds. Pick up a fold in each hand and walk forward approximately 25 feet (8 m) as the first person moves forward the same distance (Figure 4.103).

Step 7: Each person moves to pick up hose in the same manner until the hose is completely picked up (Figure 4.104).

Figure 4.103 STEP 6: Walk forward approximately 25 feet (8 m) so that a second person can pick up a fold in each hand.

Figure 4.104 STEP 7: Both firefighters walk forward approximately 25 feet (8 m) so that the third firefighter can pick up a fold in each hand.

Figure 4.105 STEP 8: Maintaining the 25-foot (8 m) separation, walk to your destination and progressively lay out the hose from the rearmost carrier forward.

Step 8: Maintain the 25-foot (8 m) separation and walk toward the destination. The rearmost carrier drops the last fold as resistance is felt in the trailing hose (Figure 4.105). Drop succeeding folds in the same manner from the rear of the group.

ADVANCING HOSELINES

Once hoselines have been laid out and connected for fire fighting, they must be advanced into final position for applying water on the fire. This task is best accomplished before the hose is charged because water adds considerable weight and makes the lines less maneuverable.

There are several ways for one person to advance an uncharged hoseline with a nozzle attached. In most cases, one method is as good as another, but the important thing to remember is that the nozzle should be secured so that it will not be dropped. This means that hose should not be advanced by simply holding onto the nozzle and walking forward because the hose could snag and pull the nozzle from the hands. A more secure method is to drape the hose over one shoulder with the nozzle in front. Holding the nozzle on the opposite side of the body will increase stability (Figure 4.106). If it

Figure 4.106 Hold the nozzle across the body to increase stability.

is necessary to have the hands free, as when climbing a ladder, the nozzle should hang over the shoulder and rest on the back, as shown in Figure 4.107.

Figure 4.107 A hands-free carry is accomplished by hanging the nozzle over the back.

Advancing a charged line is more difficult than advancing an uncharged line because hose becomes stiff, heavy, and unwieldy when filled with water and pressurized. Nevertheless, moving a charged hoseline is often necessary, as when a hoseline is advanced to the door of a structure, the nozzle opened to knock down fire, and then the hose is moved forward to extinguish fire deeper into the building.

As with an uncharged hose, the nozzle should always be protected. More important, however, is the need to maintain control of the nozzle at all times. This means that before changing positions, the nozzle should be closed to eliminate the nozzle reaction that makes the hose difficult to control.

It is always recommended that at least two persons be assigned to every charged hoseline to make control of the nozzle and hose easier during fire fighting. This practice is also consistent with the "buddy" system, in which no person works alone in a dangerous environment. When it becomes necessary, however, for one person to advance a charged hoseline, one of the best methods is to place one arm under the hose and grasp the opposite arm to lock the hoseline into place. The opposite hand should always be on the nozzle shutoff to maintain control of the water flow (Figure 4.108).

Figure 4.108 To advance a charged hoseline single-handed, place one arm under the hose and grasp the opposite arm. Keep the opposite hand on the shutoff.

Another method for advancing hose without help is to attach a hose strap or rope tool to the hose and use it to pull the hose forward with the body. Attach the strap to a point on the hose so that when the loop is placed over the shoulder the nozzle is within easy reach (Figure 4.109).

When two people advance a hose, similar methods can be used to control the hose. The nozzle firefighter can use either the arm-lock method or the hose-strap method. A second firefighter on the same side of the hose, pulls the hose forward with the hands or with a hose strap (Figure 4.110). If other persons are needed to pull the hose (as for the larger sizes of hose), they should space themselves on the same side of the hose at 10-foot (3 m) intervals and pull the hose in the same manner.
NOTE: It is not absolutely necessary that the first

and second firefighters stand on the same side of the hose, but this does make it easier for them to move through a doorway.

Figure 4.109 Attach a hose strap or rope tool so that the nozzle is within easy reach when pulling the hose forward with the body.

Figure 4.110 The nozzle firefighter directs the stream. A second firefighter can be used to pull the hose forward.

Advancing Hose Up a Stairway

Advancing a hoseline up a stairway can be a difficult task because of the tendency for a charged hose and its couplings to become snagged on stairway turns. For this reason, hose should be advanced up a stairwell *before* charging the line. The accordion shoulder carry is particularly well suited for this operation because the hose is laid progressively as the firefighter climbs the stairs. The line should be laid against the outside of the wall to avoid sharp bends (Figure 4.111).

Figure 4.111 Hoseline should be laid against the outside of the stairwell wall to avoid kinks.

In some buildings the stairwells have no walls to separate the staircases. This arrangement creates an opening in the center that makes it possible to extend hose without laying it on the stairs. In this situation, the easiest way to extend hose from the upper floors to the ground floor may be to carry a bundle of hose to the required floor and then lower the hose down the well opening (Figure 4.112). A standpipe pack, described in Chapter 7, is particularly well suited for this purpose.

Figure 4.112 In open stairwells, lower the hose down the well opening.

Advancing a charged hoseline up a stairwell is a difficult task that should only be attempted for short distances. If the hose must be moved more than one floor, it may be more expedient to simply clamp the hose behind the nozzle, detach the nozzle, and attach additional uncharged hose to the end (Figure 4.113). Extend the additional hose up the staircase, attach the nozzle, and recharge the hose as soon as the nozzle is positioned for the attack. If there is an open well, a charged hose may be passed up hand-over-hand by persons standing at each landing.

It is generally not advisable to stretch more than two hoselines up a stairway because multiple hoselines impede foot traffic and tend to become entangled. If more than two hoselines are needed on upper floors, take additional lines up other staircases or up ladders and through windows.

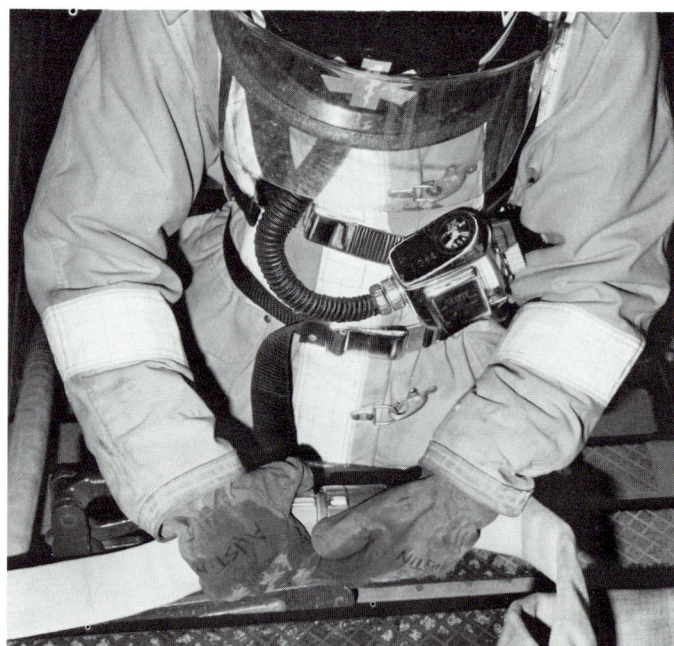

Figure 4.113 Rather than move charged hose up a stairwell, clamp the hose, detach the nozzle, and add uncharged hose to extend the line up the stairs.

Advancing Hose Up a Ladder

As with staircases, hoselines should be advanced up ladders before being charged. If a charged hose must be moved up a ladder, it is not only easier but also safer to drain the hose before advancing it up the ladder. As stated before, the person carrying the nozzle should do so with the hose draped over one shoulder so that the nozzle hangs at the back. This frees both hands for climbing the ladder. Space additional persons 10 feet (3 m) apart with each carrying the hose using a strap over the shoulder so that hands are free for climbing (Figure 4.114). If straps are not available, each person should loop the hose on the shoulder so that the hands are free (Figure 4.115). If the group wishes to advance the hose without moving the handlers up the ladder, each person should lock in and simply pass the hose along hand-over-hand.

If it becomes necessary to move charged hose up a ladder, more people are required to handle the combined weight of the hose and water. Keep in mind, however, that ladder loading will increase because of the weight of the charged hoseline. If the hose is of a larger size, increase spacing between handlers to more than the standard 10 feet (3 m). All persons on the ladder should lock in and pass the hose upward hand-over-hand. While it is possi-

Figure 4.114 Use a hose strap to free both hands for climbing.

Figure 4.115 Loop the hose over the shoulder to free both hands for climbing.

ble to carry a charged hose up a ladder using hose straps, this is not recommended because it is difficult to coordinate each person's rate of climb. Faster climbers may push the hose ahead and cause slower climbers to lose balance or to lose control of the hose.

Advancing a Booster Line

Booster hose is usually stored on reels and, unlike other hose, is at least partially charged with water when in storage on the reels. For this reason, booster hose can be fairly heavy when fully extended. A booster hose reel has an adjustable brake that prevents the reel from inadvertently turning when the apparatus is in motion. It also prevents the reel from spinning after the hose is pulled, which causes the remaining hose to uncoil on the reel. If the brake is applied too firmly against the reel, pulling the booster hose becomes difficult.

To advance booster hose directly from the reel, simply grasp the nozzle firmly in both hands and walk toward your destination. If moving a long distance, it may be necessary for other persons to assist by pulling the hose some distance behind the person at this nozzle (Figure 4.116). This distance varies with the length of hose, distance pulled, and obstacles encountered en route to the destination. It also helps to have a person at the apparatus pulling hose from the reel. If the hose reel brake is set too tightly, it should be adjusted to apply less friction, thus easing the task of pulling hose (Figure 4.117).

OPERATING HOSELINES

The methods of applying water through a nozzle attached to an attack hose vary with the size of hose used, the fire problem, and with the type of attack (see IFSTA **Essentials of Fire Fighting**, "Fire Suppression Techniques"). An important point to remember with all methods, however, is that the nozzle must be controlled at all times. This means that the operator should keep one hand on the shutoff valve whenever water is flowing through the nozzle (Figure 4.118). This will permit immediate shutdown of the nozzle if the hoseline becomes uncontrollable because of a pressure surge, loss of balance, or other problem.

Figure 4.116 When moving booster hose long distances, other firefighters can assist by pulling the hose at regular intervals along its length.

Figure 4.117 Adjust the brake on the reel to ease the task of pulling the hose.

Figure 4.118 One hand should be kept on the shutoff valve of the nozzle whenever water is flowing. *Courtesy of Greg Colby.*

One-Person Methods

When one person is required to work a nozzle unassisted, some means must be provided for bracing the hoseline. Position the hose so that it is extended straight back for at least 10 feet (3 m) behind the nozzle. Stand facing the objective with the feet spread at least shoulder width apart for good balance. Grasp the hose directly behind the nozzle with one hand and the shutoff with the opposite hand. The hose may be further anchored by placing the foot upon the hose.

Water flowing through any hoseline larger than 1 inch (25 mm) in diameter usually causes significant nozzle reaction, which is a force that pushes the hose backward. Use a hose strap to gain additional control of the hose. Attach the strap to the hose, then place the strap over the shoulder so that the body absorbs the reaction force of the nozzle. Figure 4.119 shows two ways to attach and use a hose strap.

If you need to move the hose during the attack operation, close the nozzle and move the hose to the new location. Then straighten the hose behind and operate the nozzle in the same manner as before.

Another means of controlling a hoseline when working alone is to reduce the water flow with the nozzle shutoff valve. Reducing the water flow also reduces the nozzle reaction, thus making the hose easier to handle. If you are using an adjustable flow nozzle, you may also select a lower flow setting to accomplish the same purpose.

The combination attack method of applying water to a fire requires rotating a nozzle set to a fog stream in a clockwise "O" pattern. If this attack is done through a window opening, it is usually easier to open the nozzle, adjust the pattern, and then grasp the hose approximately 2 feet (0.6 m) behind the nozzle with both hands. Holding the hose firmly, reach inside the window with the nozzle and rotate the nozzle clockwise several times in a whipping motion (Figure 4.120). Withdraw the hose and close the nozzle to observe the extinguishing effect of the applied water. Keep in mind that use of this "whip" method results in less direct control of the shutoff valve because both hands are on the hose. *Maintain a firm grip on the hose at all times throughout the operation.*

Figure 4.119 Two ways to use a hose strap to counteract the backward force from the nozzle.

Figure 4.120 Whip the nozzle clockwise in a circular motion to apply a fog stream of water.

When a large hoseline is positioned outside a structure for extended periods (for example, an exposure protection line), one person can safely and effectively operate the nozzle alone if the hose is properly arranged. Take approximately 25 feet (8 m) of the hose immediately behind the nozzle and form a loop. Pass the nozzle beneath the loop so that the loop rests on the end of the hose approximately 2 feet (0.6 m) behind the nozzle. This arrangement allows upward and downward movement of the stream and some movement side to side. Secure the loop by tying the hose at the crossover point with a hose strap. Kneel or sit on the hoseline at the crossover point and operate the nozzle (Figure 4.121).

Figure 4.122 Make a hoseline more secure by strapping it to a fixed object.

Figure 4.121 Pass the nozzle beneath the loop so the loop rests on the hose, then sit or kneel on the crossover point. *Courtesy of Dan Tinnel.*

One person can also control a hoseline for similar purposes if it is securely tied to a fixed object such as a telephone pole or parking meter (Figure 4.122).

Two-Person Methods

As stated earlier, the preferred practice when operating any hoseline during a fire attack operation is to have at least two firefighters on the line. This provides increased safety in case one of the firefighters becomes disabled. It also enables the firefighters to make use of the full waterflow potential of the nozzle.

The person at the nozzle holds it in much the same manner as when working alone — one hand on the hose directly behind the nozzle and the other hand on the nozzle shutoff valve. The nozzle person does not, however, brace the hose with the leg. The responsibility for anchoring the hose is given to the backup person so that the nozzle person has more freedom of movement than if working alone. The backup person stands on the same side of the hose and grasps it with both hands. This person cradles the hose against the inside of the closest leg and braces the hose against the front of the body and hip (Figure 4.123). The backup person thus takes most of the backward force of the nozzle.

Basic Methods of Handling Hose 133

Figure 4.123 The backup firefighter braces the hose against the front of the body and hip.

As in the one-person method, two people can better control the hose by using hose straps. Each person should attach hose straps and loop their respective straps in such a manner that the backward force from the nozzle is shared by each person (Figure 4.124).

Figure 4.124 Attach hose straps so the backward force from the water is shared by both firefighters.

Another method may be used when a nozzle is equipped with handles. One person grips the shutoff with one hand and the handle with the other hand. The other person grasps the near handle with both hands (Figure 4.125). Hose straps increase stability.

Another simple method for operating a hoseline with two persons involves the use of a straight bar about 4 feet (1 m) long. Secure the bar to the hoseline directly behind the nozzle with a hose strap. Each person stands on either side of the nozzle with the bar placed against the front of the body in the hip area. One person grasps the nozzle shutoff valve with one hand and the bar with the

Figure 4.125 Grasp the handles to provide stability.

opposite hand. The other person grasps the hose with one hand and the bar with the opposite hand. Both persons lean into the bar to counteract the back pressure. If the line must be advanced, both persons simply walk forward against the bar (Figure 4.126).

to 2.4 m) behind the nozzle. The backup persons handle the bar as in the two-person method, leaning into it with their bodies to take the back pressure from the stream. The person at the nozzle operates the shutoff and directs the stream (Figure 4.127).

Figure 4.126 When using a hose bar, simply walk forward against the bar to advance the line.

Figure 4.127 Backup firefighters use the bar to absorb the nozzle reaction while the firefighter at the nozzle directs the stream.

Three-Person Methods

There are a number of methods for three persons to hold and operate a charged line. One method involves a single person at the nozzle backed up by two persons using a straight bar attached to the hoseline with a hose strap. The bar is attached to the hose approximately 6 to 8 feet (1.8 m

A second method uses three persons in a line staggered on opposite sides of the hose. The first and third persons stand on one side of the hose, the second person stands between them on the opposite side. Spacing is the same as for several persons advancing a hose: second person 4 feet (1 m) back; third person 10 feet (3 m) farther back. Hose straps provide additional stability (Figure 4.128).

Figure 4.128 Place firefighters along the hoseline with the hose straps to provide additional stability.

Rolling a Loop to Advance a Charged Hoseline

When it is necessary to move the nozzle forward a short distance after the line has been fully extended, there is usually no way to do this except to add a section of hose at the nozzle end. This takes considerable time because it requires shutting down the line. There is a simple method, however, for advancing the nozzle a short distance if the hose has been laid with curves in the line. A hoseline that is not absolutely straight will provide slack that can be used to move the nozzle forward. This can be done easily by simply straightening the hose progressively from the water source and moving toward the nozzle. As the slack hose accumulates, it will tend to form an "S" shape. Lay one segment of the "S" over to form a large loop, then stand the loop upright and roll the slack hose toward the nozzle in a manner similar to rolling a hoop (Figure 4.129). As the slack is removed, the hose will lay out in a straight line and the nozzle can be moved forward.

Figure 4.129 Lay one segment of the "S" over the other to form a large loop, then roll it toward the nozzle to advance the slack hose.

Chapter 4 Review

Answers on page 235

MULTIPLE CHOICE: Circle the correct answer.

1. The main advantage of using the butterfly method of unrolling a straight roll is that _____.
    A. the hose can be laid out faster
    B. the inside coupling will not be damaged
    C. it can be accomplished by one person
    D. it is well suited for interconnected hose lengths

2. Which carry is especially well suited for moving 3-inch (77 mm) hose up the stairs of a high-rise?
    A. Shoulder loop
    B. Horseshoe shoulder
    C. Horseshoe underarm
    D. Modified accordion shoulder

3. When it is necessary for one person to drag a section of hose, the person should drag the hose by grasping _____.
    A. the male coupling
    B. the female coupling
    C. both couplings
    D. the folded hose end

TRUE-FALSE: Mark each statement true or false. If false, explain why.

4. Sections of fire hose should be connected so that edges are in the same plane.
    ☐ T ☐ F _____

5. Most fire nozzles have female threads.
    ☐ T ☐ F _____

6. When making a straight roll, the coil should be open enough at the center to allow the fingers to be inserted.
    ☐ T ☐ F _____

7. The twin donut roll is especially well suited for LDH.
    ☐ T ☐ F _____

8. The shoulder loop carry begins with the carrier placing one end of the hose so that the coupling is behind the body at waist height.
    ☐ T ☐ F _____

Basic Methods of Handling Hose **137**

9. It is better to advance a dry hoseline rather than a charged line.
   ☐ T ☐ F _____

10. The recommended method of advancing a hoseline is to grasp the nozzle and walk forward with the hose behind.
    ☐ T ☐ F _____

11. When operating a charged hoseline, it is best to close the nozzle before changing positions.
    ☐ T ☐ F _____

12. When a dry hoseline is advanced up a ladder, the hose should be draped over one shoulder with the nozzle hanging in front.
    ☐ T ☐ F _____

13. The recommended method for advancing a charged hoseline up a ladder is for several firefighters to lock in and pass the hose upward hand-over-hand.
    ☐ T ☐ F _____

14. Use of hose straps is recommended when one person is operating a hoseline but are not necessary when two or more persons operate a line.
    ☐ T ☐ F _____

15. When three persons are operating a charged line, they should all be positioned on the same side of the hose.
    ☐ T ☐ F _____

SELECT: Circle the correct response.

When using the knee-press method to break a tight screw connection, work with the hose on the ground. First grasp the hose behind the **16.** (male, female) coupling, then stand the connection on end with the **17.** (male, female) coupling below. The hose should be bent **18.** (close to, some distance from) this coupling. Place one knee on the hose and shank of the **19.** (male, female) coupling, apply weight to the coupling, and snap the swivel in a **20.** (clockwise, counterclockwise) direction.

When attaching a female hose coupling to a fixed male fitting, align the coupling with the fitting and turn the swivel **21.** (clockwise, counterclockwise). When attaching a male hose coupling to a fixed female fitting, align the coupling with the fitting and turn the swivel **22.** (clockwise, counterclockwise).

138 HOSE

FILL IN THE BLANK: Fill in the blanks with the correct response.

23. The bundle in the modified shoulder carry has _____ folds of equal length.

24. When several persons are used to advance a hoseline, the second person should be _____ foot/feet (_____ m) behind the first and the others should be at _____ foot/feet (_____ m) intervals.

SHORT ANSWER: Answer each item briefly.

25. It is recommended that hose connections on a fireground be made hand tight rather than using spanners. Why?

26. You noticed that a coupling on the fireground continues leaking even after you have tightened it as much as possible by hand. What should you suspect is the problem?

27. In which one-person method(s) of connecting hose should the male coupling be upturned?

28. When two persons are connecting Storz couplings, one person aligns the couplings, inserts the lugs and turns the swivel to engage the lugs. What should the other person be doing during this operation?

29. When making a straight roll of hose with screw-thread couplings, which coupling should be on the outside of the roll?

30. With regard to couplings, what is the major difference between the straight roll and the donut roll?

31. Sections of hose are moved by carrying or dragging. Which method is preferable? Why?

32. When one person is operating a hoseline, where should the hands be placed?

Basic Methods of Handling Hose **139**

33. When two or more persons are operating a hoseline, where should the nozzle operator's hands be placed?

LISTING

34. List the three one-person methods of connecting hose.

    A. _____
    B. _____
    C. _____

35. List the three methods of breaking a tight screw-thread connection.

    A. _____
    B. _____
    C. _____

36. List the two methods that can be used to secure a large exposure protection line so it can be operated safely by one person for an extended period of time.

    A. _____
    B. _____

DISCUSSION QUESTIONS

Which one-person method do you feel is best for connecting hose? Why? If you feel that different methods are preferable at different times, explain why.

Which method of rolling hose do you feel is the most useful? Why?

LEARNING ACTIVITY

Practice the following hose carries and rolls. Work alone or with others, according to each method.

Shoulder Loop Carry	Straight Roll
Accordion Shoulder Carry	Donut Roll
Modified Accordion Shoulder Carry	Twin Donut Roll
Horseshoe Carry	Self-Locking Twin Donut Roll
Horseshoe Underarm Carry	

5

Supply Hose Loads and Layout Procedures

142 HOSE

This chapter provides information that addresses performance objectives described in NFPA 1001, *Fire Fighter Professional Qualifications* (1987), particularly those referenced in the following sections:

Fire Fighter I

3-13 Fire Hose, Nozzles, and Appliances.

3-13.1

3-13.2

3-13.3

3-13.6

Fire Fighter II

4-13 Fire Hose, Nozzles, and Appliances.

4-13.4

4-13.5

Chapter 5
Supply Hose Loads and Layout Procedures

Hose is used in two basic ways: in a *supply* capacity it transports water from the source to the pump; in an *attack* capacity it transports water from the pump to the nozzle(s) (Figure 5.1). The next two chapters describe a number of methods for loading an apparatus so that hose can be laid out in an expedient manner for these purposes. The hose loads and methods described in the following pages are used in some versions throughout North America. They can be adapted to many types of apparatus and hose bed configurations.

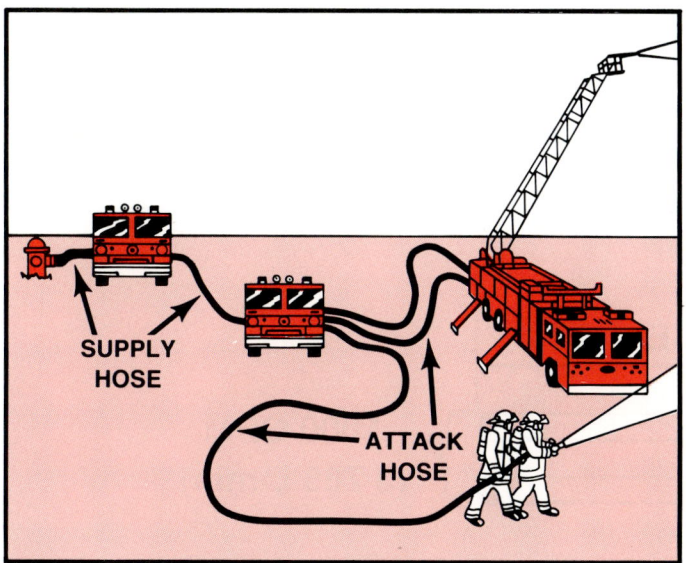

Figure 5.1 Supply hose transports water from the source to the pump; attack hose transports water from the pump to the nozzle(s).

DETERMINING WHICH HOSE LOAD TO USE

Choosing hose loads for an apparatus depends on the particular needs of the fire department. A hose load and layout procedure that works for one department may not work as well for another because conditions vary from community to community. Hose loads must be designed to accommodate the conditions within each jurisdiction.

One of the first steps in deciding which hose load to use is to define the mission of the apparatus and its crew. This is a primary factor in selecting not only the hose load, but also in determining the layout methods that will best suit the most demanding situations the unit will encounter. Loads and layout procedures should be designed so that hose can be put into service in the shortest time, with the fewest number of personnel possible.

Loads and layout procedures should also be designed for a maximum number of layout options to meet the water supply needs in any kind of situation. For example, in a small two-station fire department with one pumper at each station, the mission of the first-arriving unit is usually "initial attack." In this case, the hose might be loaded so that the pumper can lay a supply line from the hydrant to the fire as it approaches the burning structure. The second-arriving unit, which often arrives minutes later, may be designated as the "water supply" unit. This apparatus not only pumps water to the first-arriving pumper through the first-laid line but may also lay additional lines if the situation demands more water. When pumpers expect to operate in either an attack mode or a water supply mode, they should be loaded with similar hose bed configurations.

Another factor to consider when designing hose loads is the size of the apparatus pump. Pump size has a direct bearing on the manner in which hose is loaded and laid and on the size of hose used. An apparatus with a large-capacity pump should carry a hose load of sufficient size and length to deliver a volume of water equal to the maximum output of the pump. There are three basic ways to accomplish this:

- Carry large diameter hose in at least one section of the hose bed.
- Carry large amounts of small diameter hose in a split hose bed so that multiple lines can be laid.
- Carry hose on a two-piece unit: one apparatus is equipped with attack hose and the second apparatus is equipped with large amounts of supply hose.

The use of large diameter hose makes sense for a number of reasons. LDH takes up less space in the hose bed than an equivalent amount of small diameter hose. Equivalency in this case means the amount of small diameter hose needed to deliver the same amount of water the same distance and at the same pressure as a single length of large diameter hose. Small diameter hose must be laid in multiple lengths to accomplish this. For example, it takes at least five 2½-inch (65 mm) hoses to deliver 1,000 gpm (3 785 L/min) of water at the same pump pressure as a single 5-inch (125 mm) hose of the same length (Figure 5.2). Conversely, the pump pressure required to move 1,000 gpm (3 785 L/min) through two 100-foot (30 m) 2½-inch (65 mm) hoses will move the same amount of water 750 feet (230 m) through a *single* 5-inch (125 mm) hose (Figure 5.3).

Figure 5.2 It takes at least five 2½-inch (65 mm) hoses to deliver 1,000 gpm (3 785 L/min) of water at the same pump pressure as a single 5-inch (125 mm) hose of the same length.

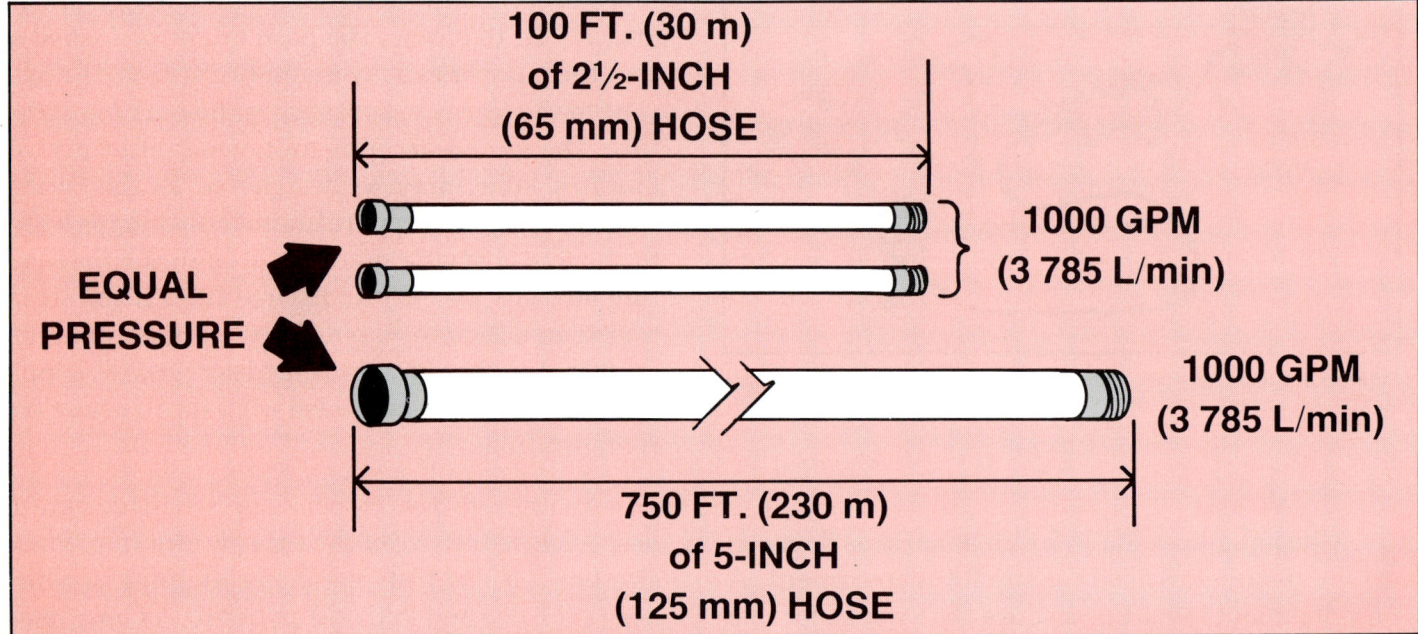

Figure 5.3 It takes approximately the same amount of pump pressure to move 1,000 gpm (3 785 L/min) through 750 feet (230 m) of 5-inch (125 mm) hose as through two 100-foot (30 m) 2½-inch (65 mm) hoses.

One of the most significant advantages of large diameter hose is that it cannot only deliver large amounts of water great distances with a relatively low loss of pressure, but that it can also deliver small amounts of water *with virtually no pressure loss*. For this reason, the fireground commander should not restrict the use of LDH to *only* large, high-volume fires.

Carrying only medium diameter hose (3-inch [77 mm] or smaller) has some advantages, however. An apparatus with a hose bed of 2½- or 3-inch (65 mm or 77 mm) hose can use the entire complement for either attack or supply. Proponents of small diameter hose argue that having the flexibility to use this hose in either capacity outweighs the disadvantage of having to lay multiple lines to supply maximum water demands. NFPA 1901, *Standard for Automotive Fire Apparatus* recommends that a minimum of 1,500 feet (450 m) of 2½-inch (65 mm) or larger hose be carried. With pumps of 1,000 gpm (3 785 L/min) and larger, this is a bare minimum unless LDH is carried because there will be insufficient hose to efficiently utilize the pump.

Carrying supply and attack hose on two separate apparatus that respond together is a practice that works well when a department has sufficient apparatus and personnel. The greatest disadvantage of such a system is its high cost, both initially and on a daily basis. The high initial outlay for a pumper makes this practice almost impossible for a fire department with limited funds.

The nature of the water source is an important factor to consider when determining hose loads. If a pumper works in a municipality that has a well-designed water system with hydrants spaced at regular intervals along each main, the need for carrying large amounts of hose is less than in a rural situation where long lays must be made from a distant water source.

Another important factor in determining the size of hose and type of load for an apparatus is whether that unit may be required to work independently for an extended time. In jurisdictions where a second-arriving pumper is more than ten minutes away, pumpers must be set up to be as self-sufficient as possible. Here again, LDH provides a means to lay a high-volume supply line with little need for another pumper to boost pressure at the hydrant.

DIRECTION OF HOSE LAYS

Before discussing the methods of loading hose, some mention must be given to the direction in which supply hose, particularly hose with threaded couplings, is laid. Threaded-coupling hose must be arranged in the hose bed so that when hose is laid, the end with the female coupling is toward the water source and the end with the male coupling is toward the fire (Figure 5.4). When hose is laid in this manner, several options are available. At the water source, hose can be connected to the male threads of a pumper discharge valve or to the male threads of a hydrant. At the opposite end, it can be connected to the auxiliary intake valve of a pumper or it can be connected directly to nozzles and appliances, all of which have female threads. There are three basic hose lays for supply hose: the forward lay, the reverse lay, and the split lay (sometimes called the combination lay).

With the *forward* lay, hose is laid from the water source to the fire. Hose beds set up for forward lays should be loaded so that the first coupl-

Figure 5.4 Threaded-coupling hose must be arranged in the hose bed so that when hose is laid, the end with the female coupling is toward the water source and the end with the male coupling is toward the fire.

ing to come off the hose bed is female (Figure 5.5). This method is often used when the water source is a hydrant and the pumper must stay at the fire location. The primary advantage with this lay is that a pumper can remain at the incident scene so that its hose, equipment, and tools can be quickly obtained if needed. The pump operator also has visual contact with the fire fighting crew and, therefore, can better react to changes in the fire operation than if the pumper is at the hydrant. A disadvantage with the lay, however, is that if a long length of medium diameter hose is laid, it may be necessary for a second pumper to boost the pressure in the line at the hydrant. This requires the use of a four-way hydrant valve so that the transition from hydrant pressure to pump pressure can be made without interrupting the flow of water in the supply hose. Another disadvantage is that one member of the crew is temporarily unavailable for a fire fighting assignment because that person must stay at the hydrant long enough to make the connection and open the hydrant.

With the *reverse* lay, hose is laid from the fire to the water source. Hose beds set up for reverse lays should be loaded so that the first coupling to come off the hose bed is male (Figure 5.6). This method is used when a pumper must first go to the fire location so that a size-up can be made before laying a supply line. It is also the most expedient way to lay hose if the apparatus that lays the hose must stay at the water source, as when drafting or boosting hydrant pressure to the supply line. A disadvantage with the reverse lay, however, is that essential fire fighting equipment, including attack hose, must be removed and placed at the fire location before the pumper can proceed to the water source. This causes some delay in the initial attack. The reverse lay also obligates one person, the pump operator, to stay with the pumper at the water source, thus preventing that person from performing other essential fire location activities.

Figure 5.6 Hose beds set up for reverse lays should be loaded so that the male coupling comes off first.

Figure 5.5 Hose beds set up for forward lays should be loaded so that the female coupling comes off first.

The split lay is a hoseline laid in part as a forward lay and in part as a reverse lay. This can be accomplished by one pumper making a forward

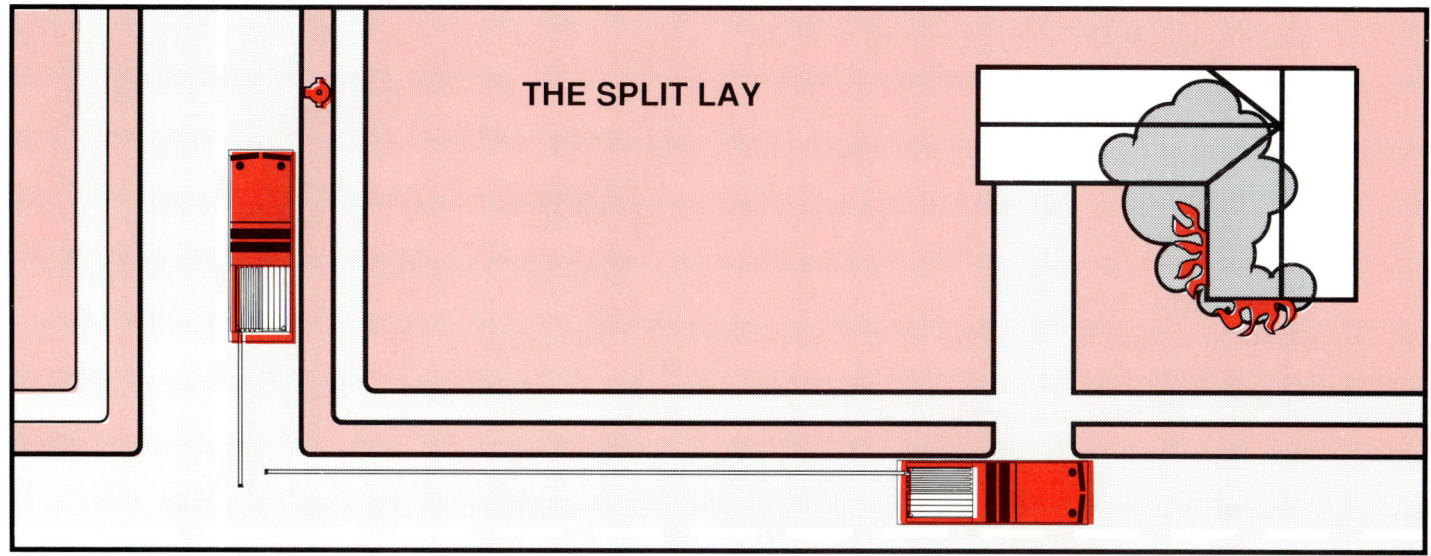

Figure 5.7 A split lay is a hoseline laid by two pumpers, one making a forward lay and one making a reverse lay from the same point.

lay from an intersection or driveway entrance toward the fire. A second pumper can then make a reverse lay to the water supply source from the point where the initial line was laid (Figure 5.7). Care must be taken to avoid making the lay too long for the pump, hose size, and required gallon per minute delivery.

HOSE LOADING GUIDELINES

When loading hose, a few basic guidelines should be followed:

- Clean up grease, oil, or other substances from the ground before laying out the hose for loading.
- Check for the presence and condition of gaskets in all swivels.
- Check all swivels to ensure that they rotate freely.
- Connect hose so that the edges are in the same plane.
- Hand-tighten all connections on screw-thread couplings — do *not* use spanners.
- Load the hose so that couplings will pull off without flipping over. This may require placing a "dutchman" in the hose (Figure 5.8).
- Load LDH with all the couplings to the front of the bed. This method saves space and makes a more uniform load.

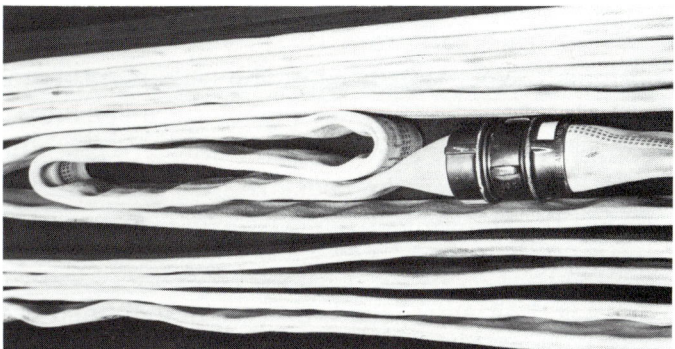

Figure 5.8 Place a "dutchman" in the hose to prevent a coupling from flipping over when the hose is pulled from the bed.

- Avoid packing hose too tightly within the bed (Figure 5.9). Overpacking causes couplings to hang up on other couplings when the hose is pulled.

Figure 5.9 Avoid packing the hose too tightly within the hose bed.

- Make a list of the section numbers as the hose is loaded. This list can be used later to identify how many sections of hose have been laid out when making hydraulics calculations.

BASIC HOSE LOADS

There are three basic hose load choices when a conventional hose bed is used: accordion, horseshoe, and flat (or a combination of these loads). Hose can also be carried in a reel load, which does not require a hose bed.

The first three loads can be used in a standard hose bed or in a bed that has been "split" with partitions. Dividing the hose bed into separate sections permits the options of laying a single line, laying lines in each direction, or concurrent laying of parallel lines. Figure 5.10 shows a typical split hose load. Note that hose in the beds is interconnected so that the entire complement can be laid in a single-length hose lay. If dual lines are laid, the couplings between the beds are first disconnected so that the couplings from the top of both beds can be pulled to start the lay.

Another way to load the hose in a divided bed is to load one side for a forward lay and the other side for a reverse lay (combination load) (Figure 5.11). This permits laying hose in either direction as required by the fire situation. When loading hose in this manner, the beds must be interconnected with a double male or double female if it becomes necessary to lay out the entire hose complement in a single-line lay.

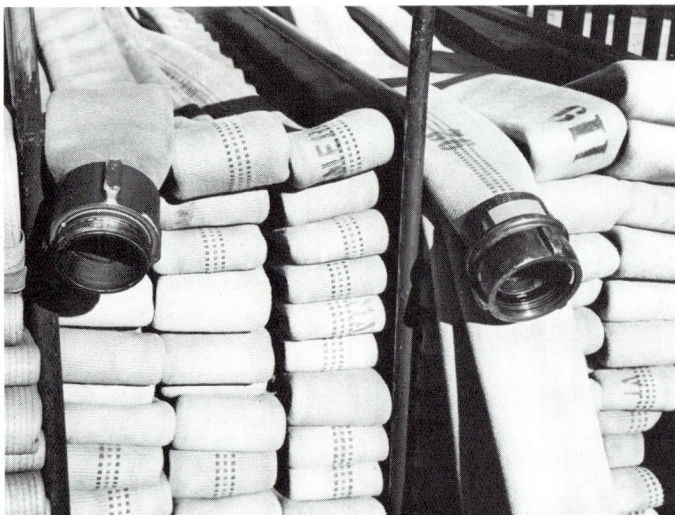

Figure 5.11 Another split hose bed load is loaded with one side set up for a forward lay and the other side set up for a reverse lay.

No matter which load is used, if the hose bed is being set up for laying hose in a forward lay, start by placing the male coupling in the hose bed. In this way, the female end of the hose will leave the hose bed first when hose is laid. For a reverse lay, start with the female coupling so that the male end of the hose leaves the bed first.

Accordion Load

The accordion load derives its name from the manner in which the hose appears after loading. The hose is laid progressively on edge in folds that lie adjacent to each other (accordionlike) (Figure 5.12). An advantage of this load is its ease of loading: its simple design requires only two or three persons (although four persons are best) and it can be loaded in a matter of minutes. Another advantage is that hose for shoulder carries can be easily taken from the load by simply picking up a number of folds and placing them on the shoulder (see Chapter 4, Accordion Shoulder Carry).

A disadvantage of this load is that the hose folds contain sharp bends, which requires that the hose be reloaded periodically if not used on a regular basis. This is so that the bends can be relocated within each length of hose to prevent damage to the lining. Another disadvantage is that the hose tends to wear along its edges. This is due to the combined effect of the vibration caused by apparatus motion and the weight of the layered hose on itself. This load is not recommended for LDH be-

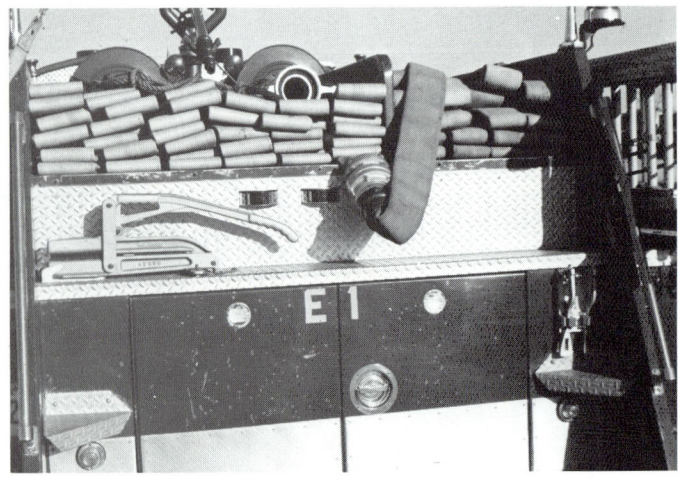

Figure 5.10 A typical split hose load with interconnected beds.

Supply Hose Loads and Layout Procedures **149**

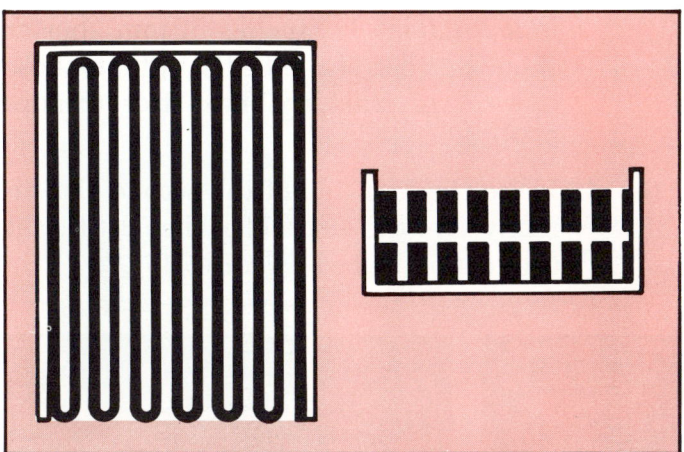

Figure 5.12 Hose in an accordion load is placed on edge in folds that lie adjacent to each other.

cause the remaining folds tend to collapse into the flat position as the hose lays out, which could cause them to become entangled.

NOTE: For reference purposes, the end of the hose bed closest to the tailboard is the *rear*, the end closest to the cab is the *front* (Figure 5.13). A layer of hose is sometimes referred to as a *tier*.

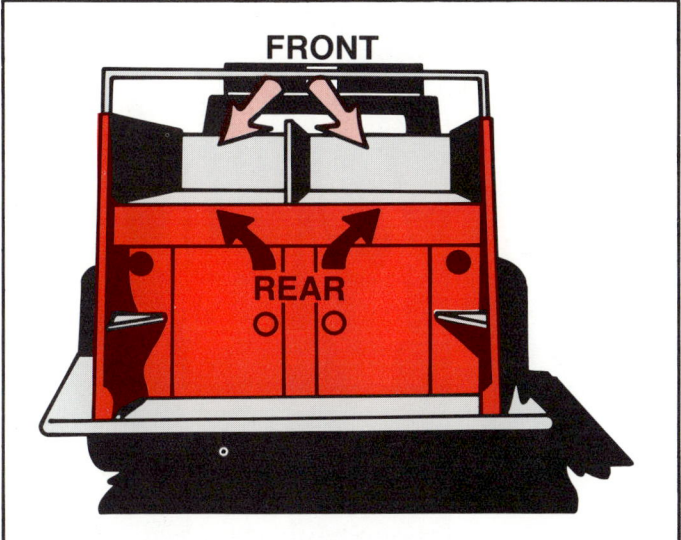

Figure 5.13 The end of the hose bed closest to the tailboard is the "rear," the end closest to the cab is the "front."

The first coupling placed in the bed should be located to the rear of the bed. It can be put to either side if the bed is not split. This procedure is for loading an accordion load into a split hose bed for a reverse lay:

Step 1: Lay the first length of hose in the bed on edge against the partition. Allow the female coupling to hang below the hose bed so that it can later be placed on top of the hose in the adjacent bed. At the front of the hose bed, fold the hose back on itself and lay it back to the rear next to the first length (Figure 5.14).

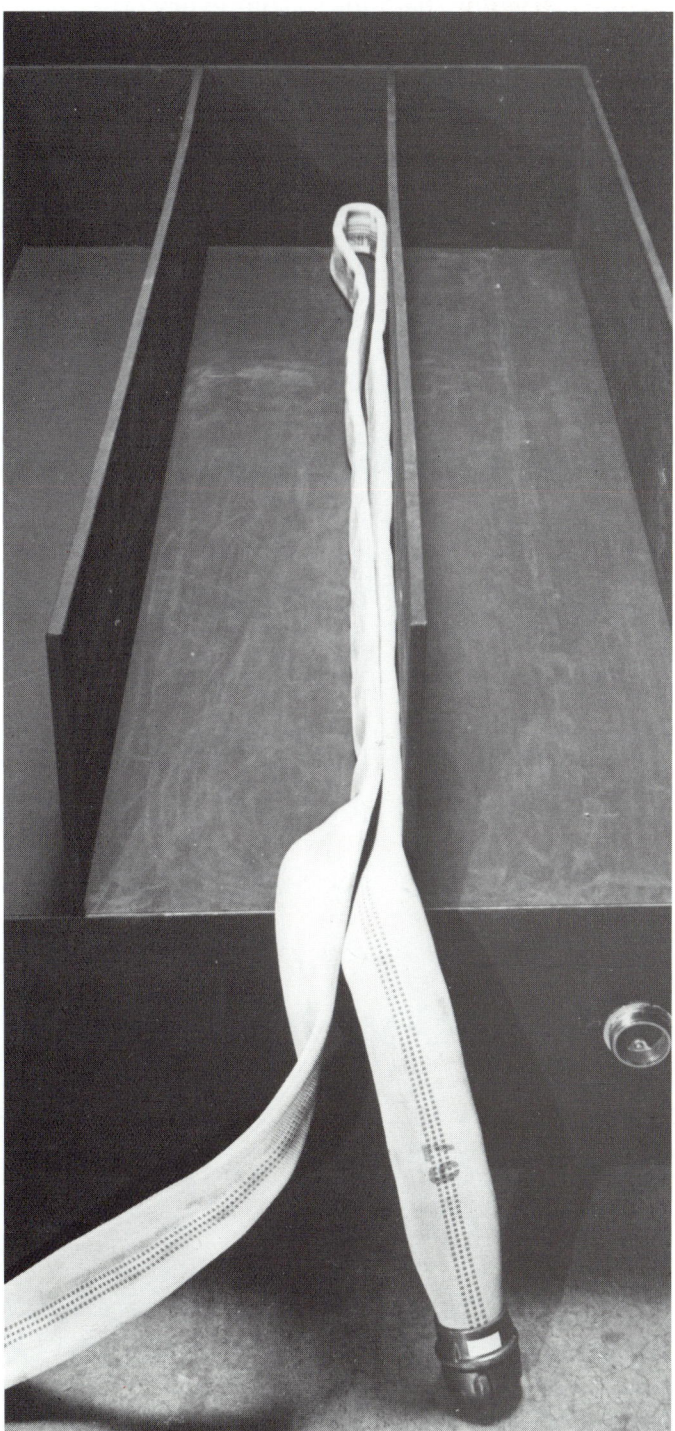

Figure 5.14 STEP 1: Lay the first length of hose into the hose bed with the female coupling hanging below the hose bed. At the front of the bed, fold the hose back onto itself and lay it back to the rear adjacent to the first length.

Step 2: At the rear of the hose bed, fold the hose so that the bend is even with the rear edge of the bed, then lay the hose back to the front (Figure 5.15).

Step 3: Continue laying the hose in folds across the hose bed. At the rear edge of the bed, stagger the folds so that every other bend is approximately 2 inches (50 mm) shorter than the edge of the bed. Angle the hose upward to start the next tier (Figure 5.16).

Step 4: Make the first fold of the second tier directly over the last fold of the first tier at the rear of the bed (Figure 5.17).

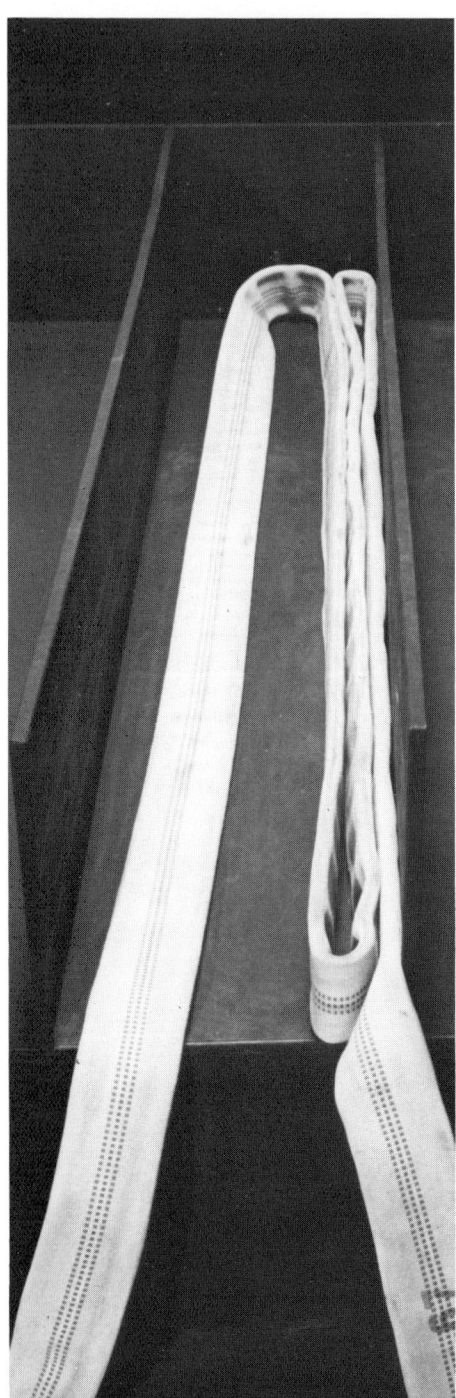

Figure 5.15 STEP 2: Fold the hose even with the rear edge of the bed and lay it back to the front.

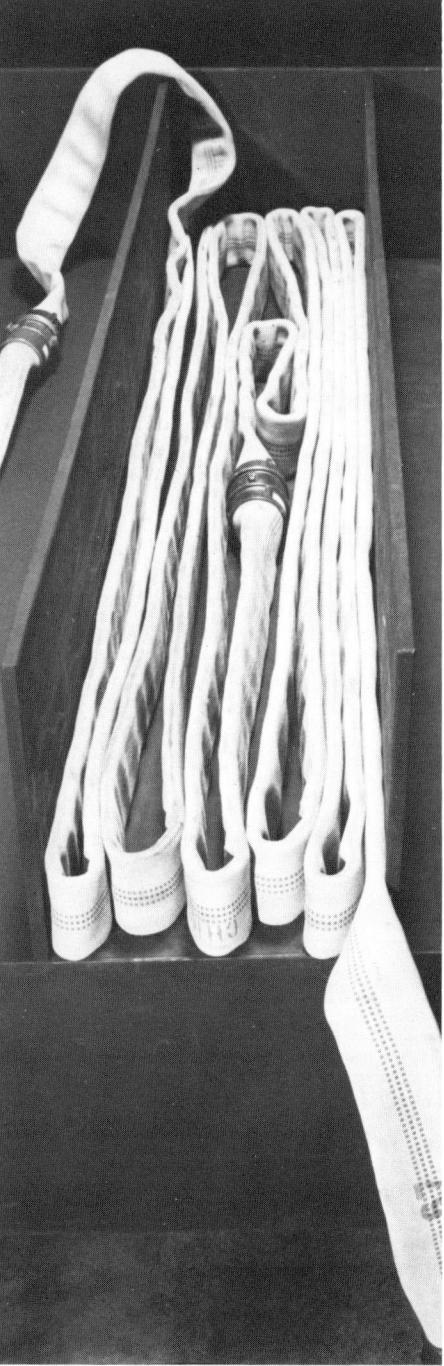

Figure 5.16 STEP 3: Continue laying the hose in folds across the hose bed. Stagger the folds so that every other bend is approximately 2 inches (50 mm) shorter than the edge of the bed.

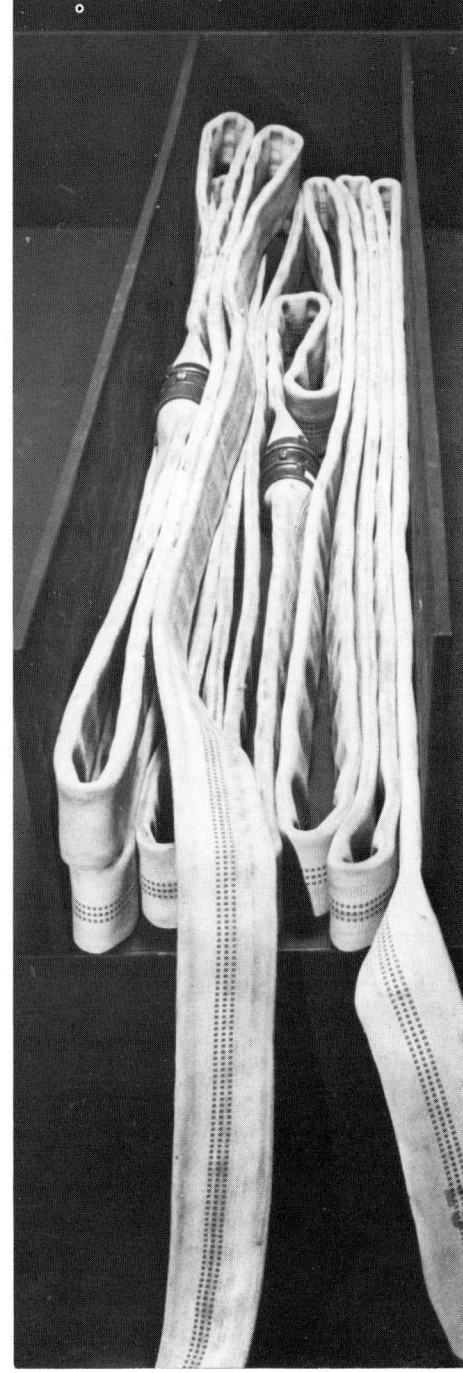

Figure 5.17 STEP 4: Place the first fold of the second tier directly over the last fold at the rear of the bed.

Supply Hose Loads and Layout Procedures **151**

Step 5: Continue with the second tier in the same manner as the first, progressively laying the hose in folds across the hose bed. Stagger the folds as before so that every other bend is approximately 2 inches (51 mm) inside adjacent bends (Figure 5.18).

Step 6: Make the third and succeeding tiers in the same manner as the first two tiers. Move to the opposite bed and load the hose in the same manner as the first side. Start with the first female coupling against the front wall of the hose bed so that it will be pulled straight from the bed when this section of hose is pulled (Figure 5.19).

Step 7: Load the second side in the same manner as the first side. When the load is completed, connect the last coupling on top with the female coupling from the first side. Lay the connected couplings on top of the hose load and pull out the slack so that the crossover loop lies tightly against the hose load (Figure 5.20).

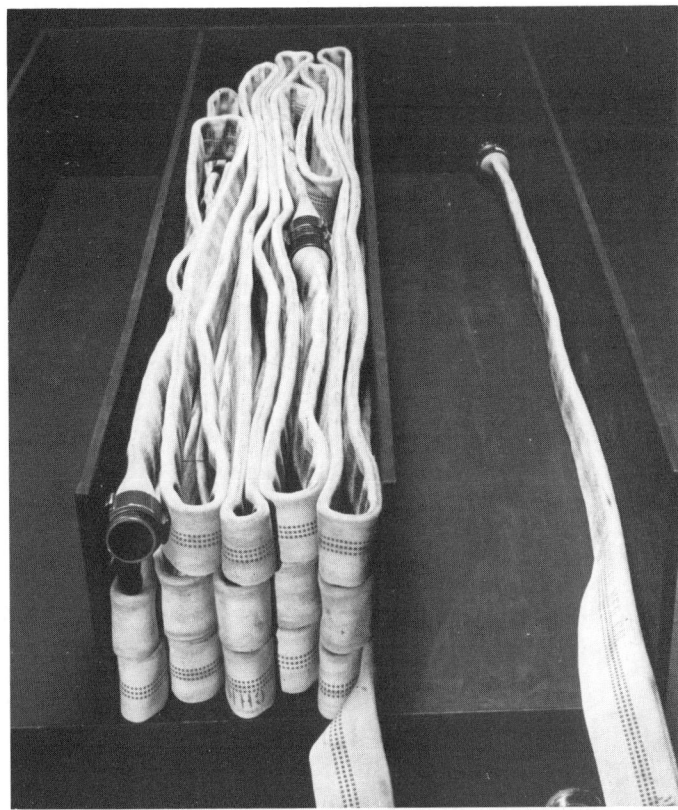

Figure 5.19 STEP 6: Load the third and succeeding tiers in the same manner as the first two tiers. Start loading the opposite bed by placing the female coupling against the front wall of the hose bed.

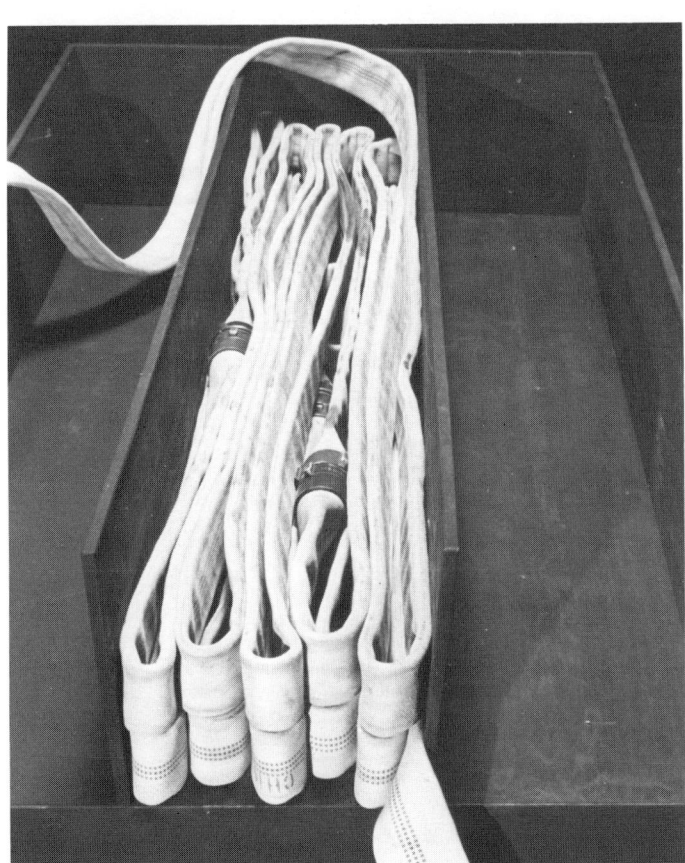

Figure 5.18 STEP 5: Continue with the second tier in the same manner as the first.

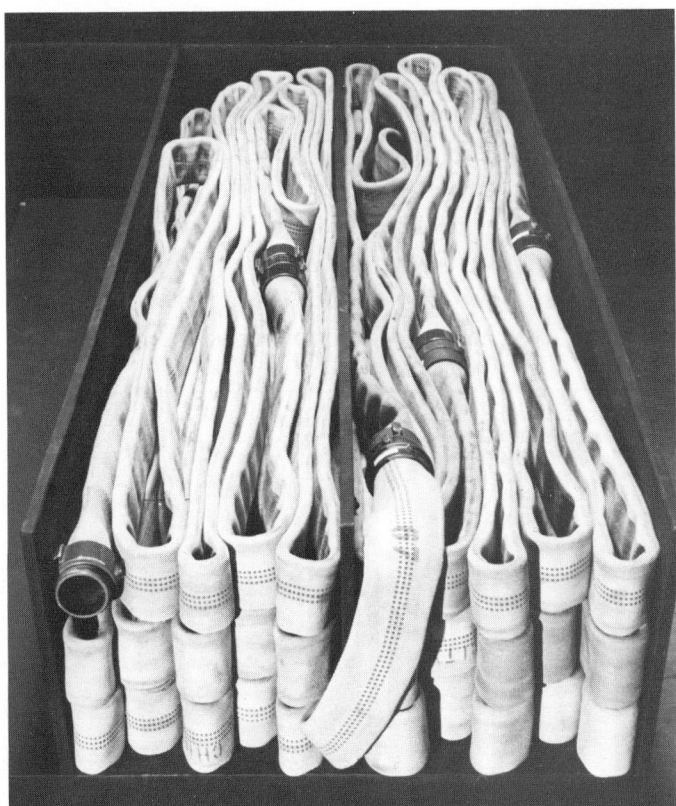

Figure 5.20 STEP 7: When the second side is completed, interconnect the beds and lay the connected couplings on top of the hose load.

Horseshoe Load

The horseshoe load is also named for the way it appears after loading — it resembles a horseshoe (Figure 5.21). Like the accordion load, it is loaded on edge, but in this case the hose is laid around the perimeter of the hose bed in a U-shaped configuration. Each length is progressively laid from the outside of the bed toward the inside so that the last length is at the center of the horseshoe. The primary advantage of the horseshoe load is that it has fewer sharp bends in the hose than the accordion or flat loads. A disadvantage of the horseshoe load occurs most often in wide hose beds. The hose sometimes comes out in a wavy, or snakelike, lay in the street as the hose is pulled alternately from one side of a bed and then the other. Another disadvantage with the horseshoe load is that folds for a shoulder carry cannot be as easily obtained as with an accordion load. In this case, two persons are required to make the shoulder folds for the carry (see Chapter 4, Accordion Shoulder Carry). As with the previous load, the hose is loaded on edge, which can promote wear on the hose edges. The horseshoe load is not recommended for LDH because as the hose pays off, the hose remaining in the bed tends to fall over, which can cause the hose to become entangled.

In a single hose bed, the horseshoe load may be started on either side. In a split hose bed, lay the first length against the partition with the coupling hanging an appropriate distance below the hose bed. Gauge this distance on the anticipated height of the completed hose load so that the coupling can be connected to the last coupling of the load on the opposite side (crossover) and laid on top of the load. This allows easy disconnection of the couplings when the load must be split to lay dual lines. With a combination load (one side loaded for a reverse lay, the other side for a forward lay), use an adapter to connect identical couplings.

The procedure that follows is for a single-bed horseshoe load set up for a reverse lay:

Step 1: Placing the female coupling in a front corner of the hose bed, lay the first length of hose on edge against the wall. Make the first fold at the rear even with the edge of the hose bed (Figure 5.22).

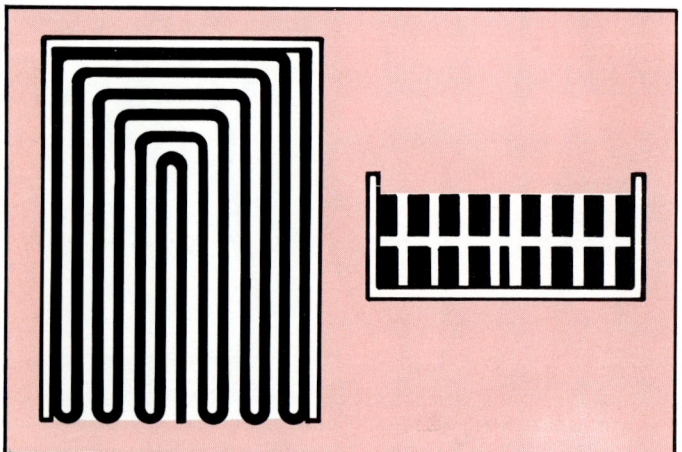

Figure 5.21 A horseshoe load resembles a horseshoe.

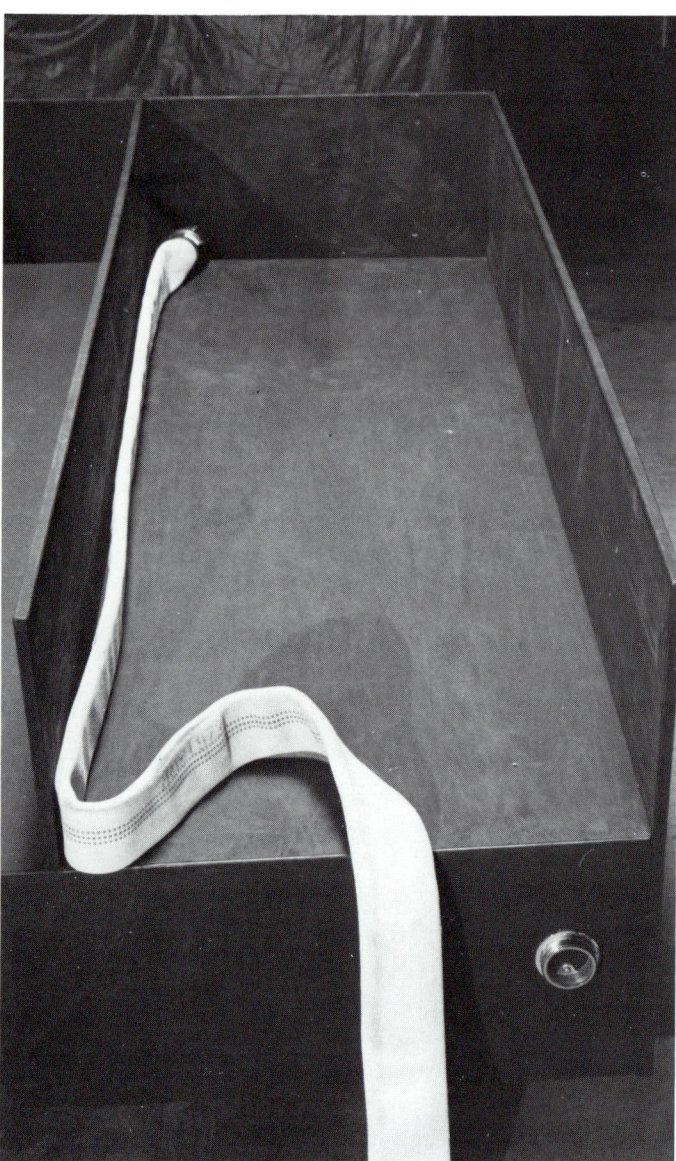

Figure 5.22 STEP 1: Lay the first length of hose on edge against the wall with the female coupling in the front corner. Make the first fold at the rear even with the edge of the hose bed.

Step 2: Lay the hose to the front, then around the perimeter of the bed so that it comes back to the rear along the opposite side (Figure 5.23).

Step 3: Make a fold at the rear in the same manner as before, then lay the hose back around the perimeter of the hose bed inside the first length of hose (Figure 5.24).

Step 4: Lay succeeding lengths progressively inward toward the center until the entire space is filled. If desired, stagger the folds so that every other bend is approximately 2 inches (51 mm) inside adjacent bends. Start the second tier by extending the hose from the last fold directly over to a front corner of the bed, laying it flat on the hose of the first tier (Figure 5.25).

Step 5: Make the second and succeeding tiers in the same manner as the first. Lay the

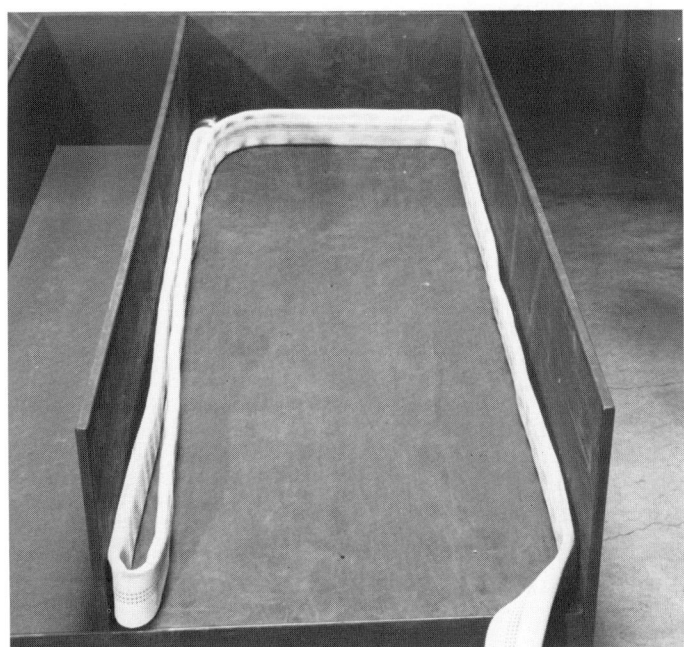

Figure 5.23 STEP 2: Lay the hose around the perimeter of the bed so that it comes back to the rear along the opposite side.

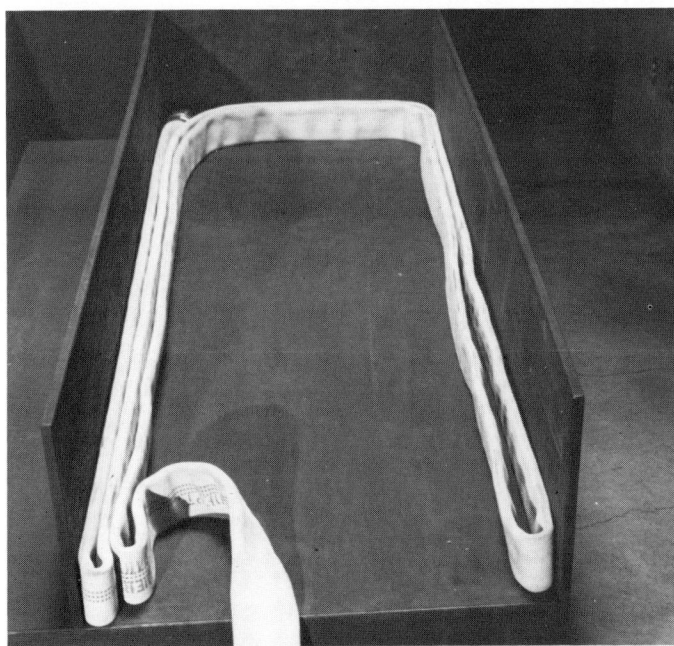

Figure 5.24 STEP 3: Lay the hose back around the perimeter of the hose bed inside the first length of hose.

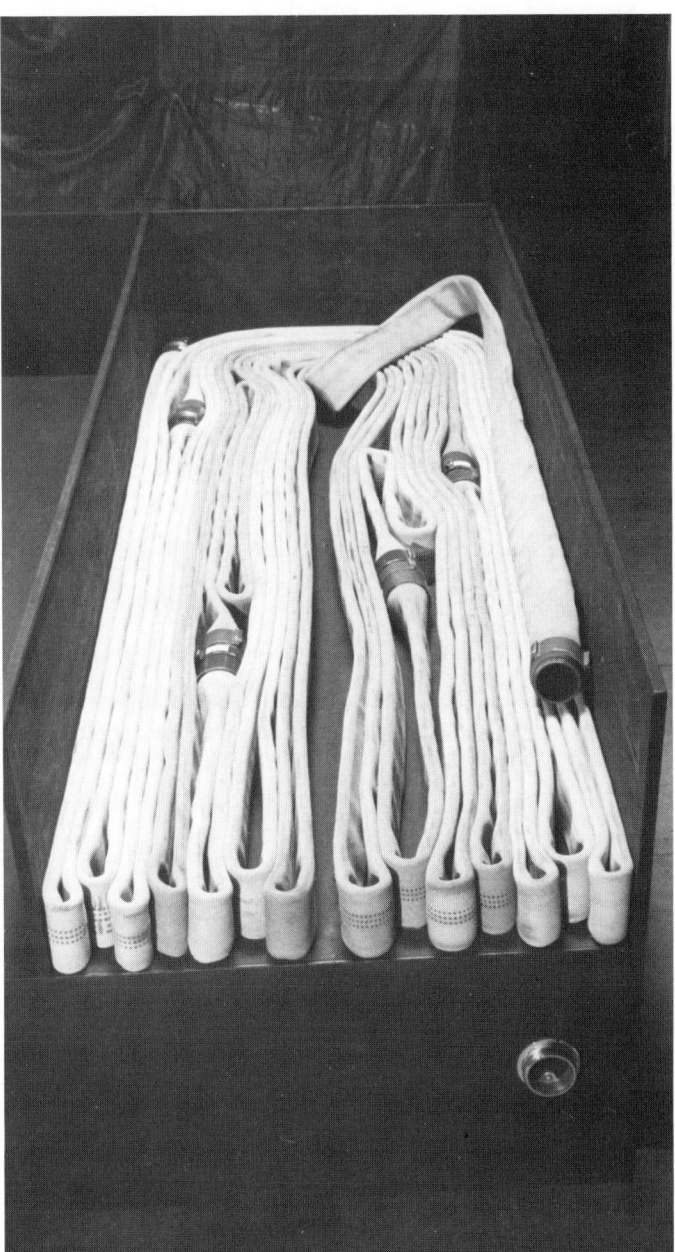

Figure 5.25 STEP 4: Lay succeeding lengths progressively inward toward the center until the entire space is filled. Lay the hose from the last fold flat over to a front corner of the bed to start the second tier.

crossover length flat on the second tier, but lay it to the opposite corner from that in the first tier (Figure 5.26). Make crossovers in succeeding tiers to alternate corners.

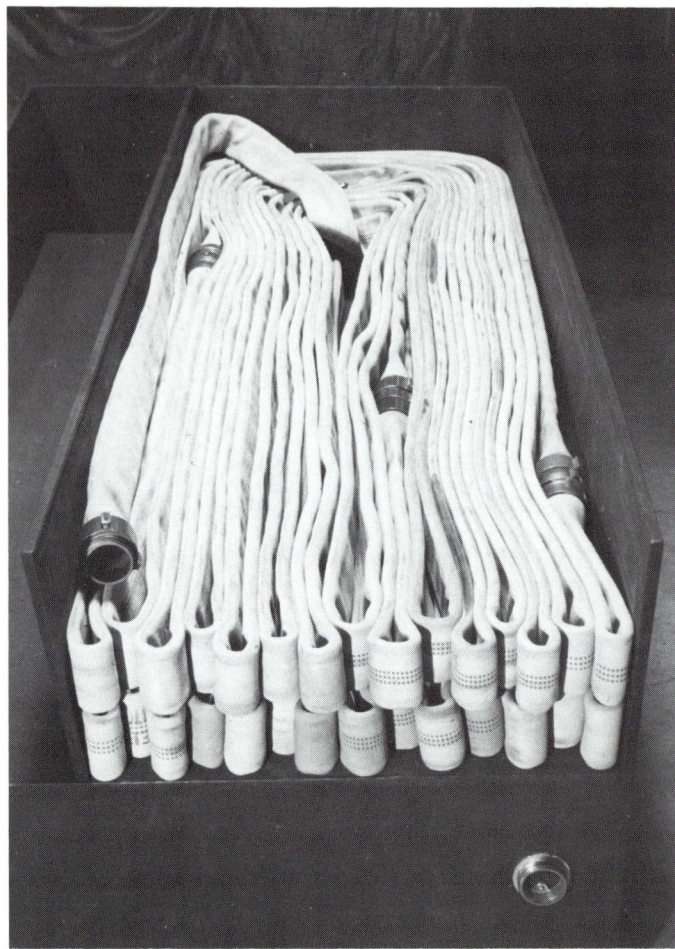

Figure 5.26 STEP 5: Load succeeding tiers in the same manner to complete the load. Make crossovers in succeeding tiers to alternate corners.

Flat Load

Of the three supply hose loads, the flat load is the easiest to load. It is suited for any size of supply hose, including LDH. As the name implies, the hose is laid so that its folds lie flat rather than on edge (Figure 5.27). Hose loaded in this manner is less subject to wear from apparatus vibration during travel. As with the accordion load, a disadvantage of this load is that the hose folds contain sharp bends, which requires that the hose be reloaded periodically to relocate bends within each length to prevent damage to the lining. Another disadvantage of the load is that it requires two persons to make folds for shoulder carries.

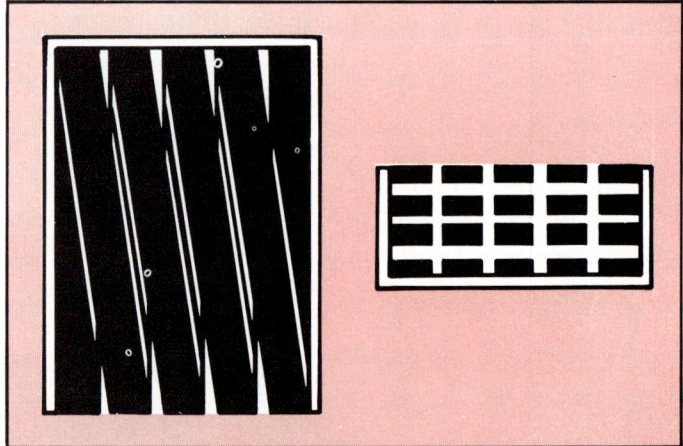

Figure 5.27 The hose in the flat load is laid flat across the bed, rather than on edge.

In a single hose bed, the flat load may be started on either side. In a split hose bed, lay the first length against the partition with the coupling hanging an appropriate distance below the hose bed. Gauge this distance on the anticipated height of the hose load so that the coupling can be connected to the last coupling of the load on the opposite side (crossover) and laid on top of the load. This allows easy disconnection of the couplings when the load must be split to lay dual lines. With a combination load, use an adapter to connect identical couplings.

The procedure that follows is for a split-bed flat load set up for a reverse lay:

Step 1: Lay the first length of hose flat in the bed against the partition with the female coupling hanging below the hose bed (to be later connected to hose in the adjacent bed) (Figure 5.28).

Step 2: Fold the hose back on itself at the front of the hose bed and lay it back to the rear on top of the previous length. At the rear of the hose bed, fold the hose so that the bend is even with the rear edge of the bed (Figure 5.29).

Step 3: Lay the hose back to the front of the bed, angling it to make the front fold adjacent to the previous fold (Figure 5.30).

Step 4: Continue to lay the hose in folds progressively across the bed to complete the first tier (Figure 5.31).

Supply Hose Loads and Layout Procedures 155

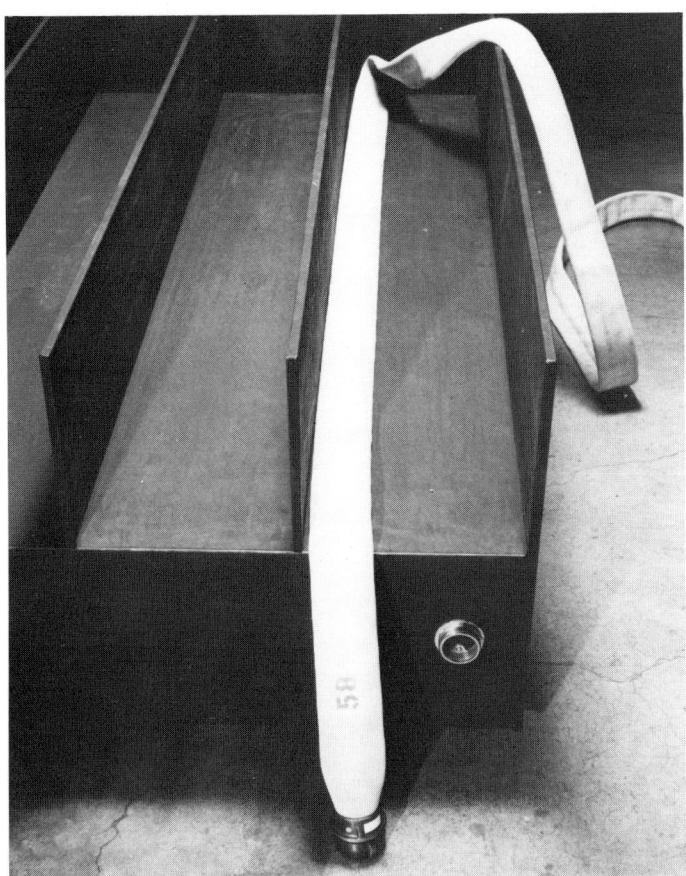

Figure 5.28 STEP 1: Lay the hose against the partition with the female coupling hanging below the hose bed.

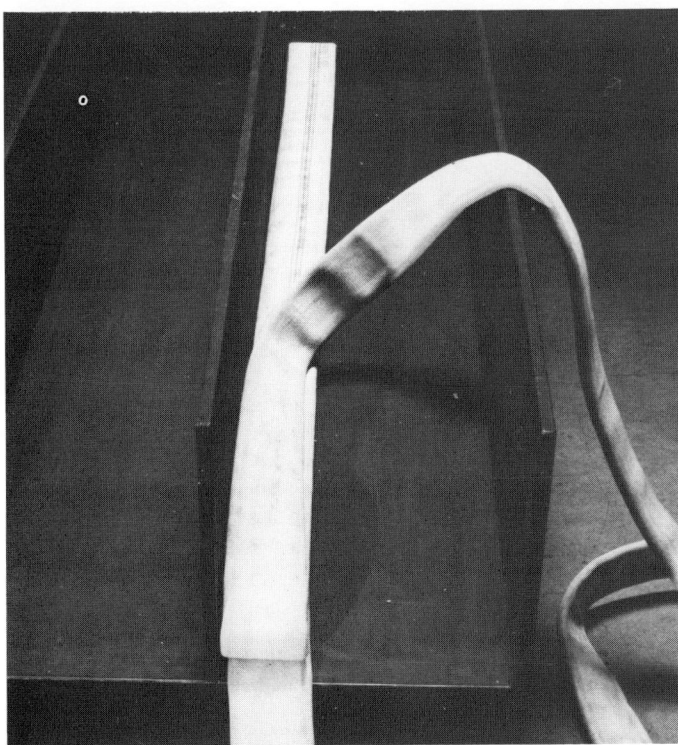

Figure 5.29 STEP 2: Make a bend in the hose at the front and lay the hose back to the rear on top of the previous length. Fold the hose even with the rear edge of the bed.

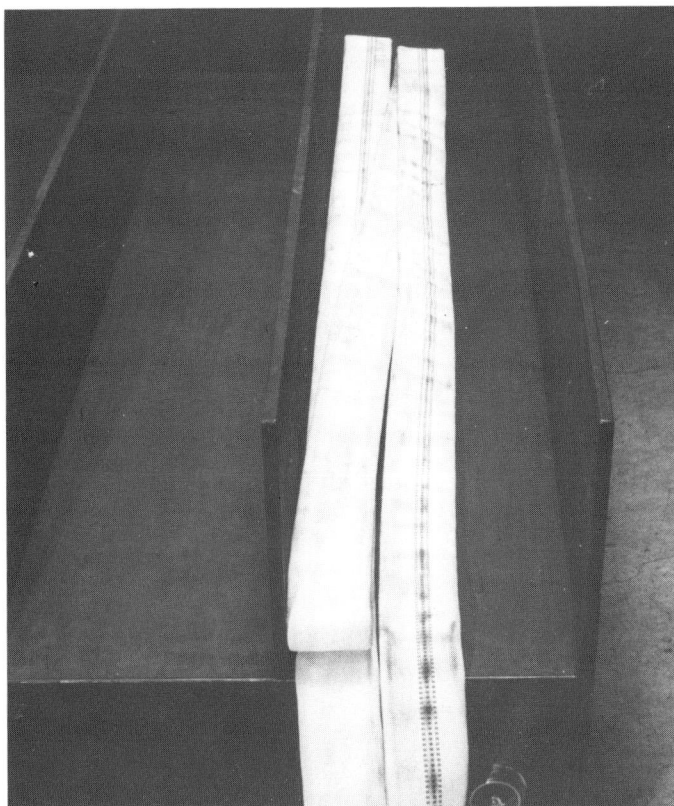

Figure 5.30 STEP 3: Lay the hose back to the front of the bed, angling it so that the front fold is made adjacent to the previous fold.

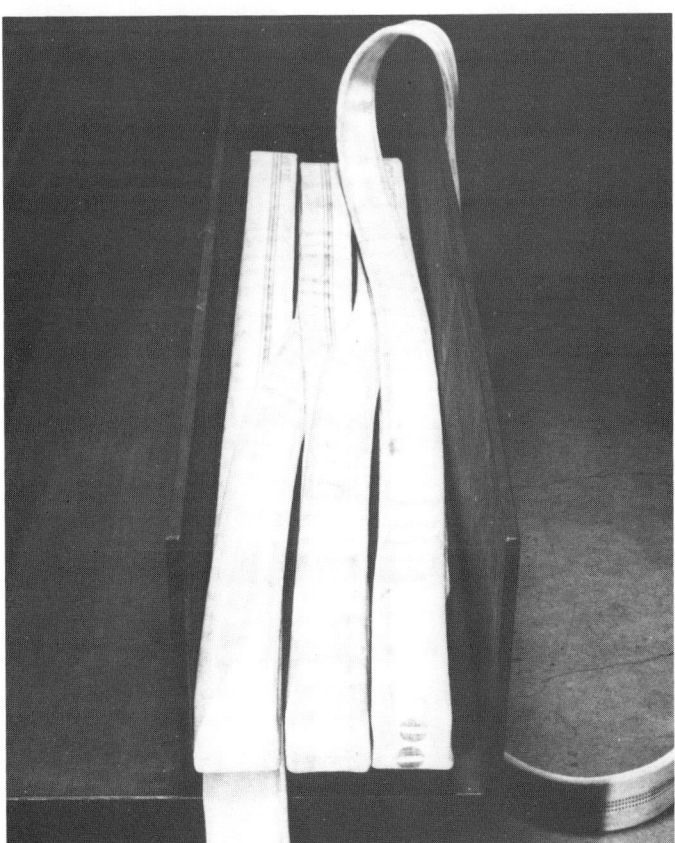

Figure 5.31 STEP 4: Continue to progressively lay the hose in folds across the bed to complete the first tier.

Step 5: Continue with the second tier in the same manner as the first, laying the hose in folds progressively across the hose bed. If desired, make the folds of the second tier approximately 2 inches (51 mm) shorter than the folds of the first tier (Figure 5.32).

Step 6: Make the third and succeeding tiers in the same manner as the first and second tiers. Align the bends of the third tier even with those of the first tier, the bends of the fourth tier with the second tier, and so on until the load is completed (Figure 5.33).

Step 7: Move to the opposite bed and load the hose in the same manner as the first side. Start the first male coupling against the front wall of the hose bed, however, so that it will be pulled straight from the bed when this last section of hose is pulled (Figure 5.34).

Step 8: When the opposite side is loaded, connect the last coupling on top with the female coupling from the first side using a double male. Lay the connected couplings on top of the hose load and pull out the slack so the crossover loop lies tightly against the hose load (Figure 5.35).

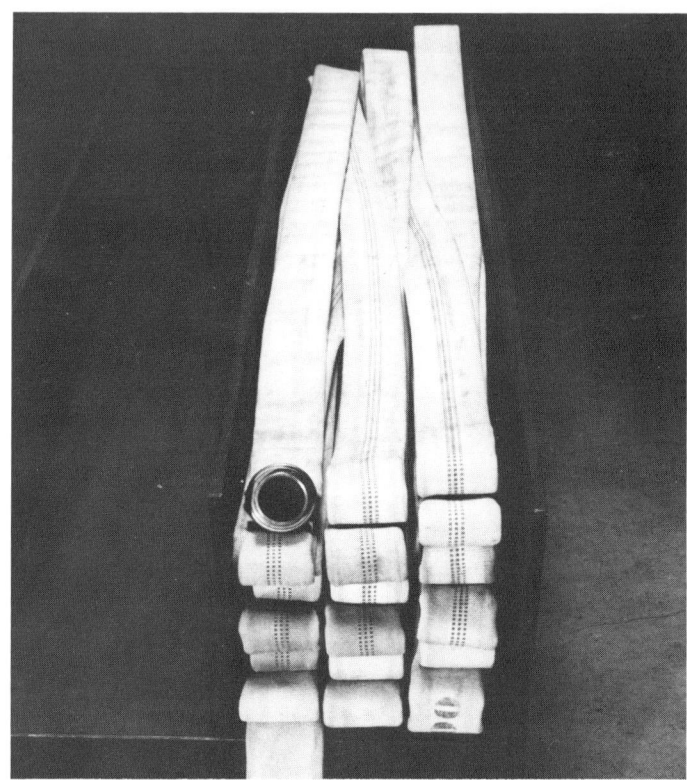

Figure 5.33 STEP 6: Continue loading succeeding tiers in the same manner as the first and second tiers.

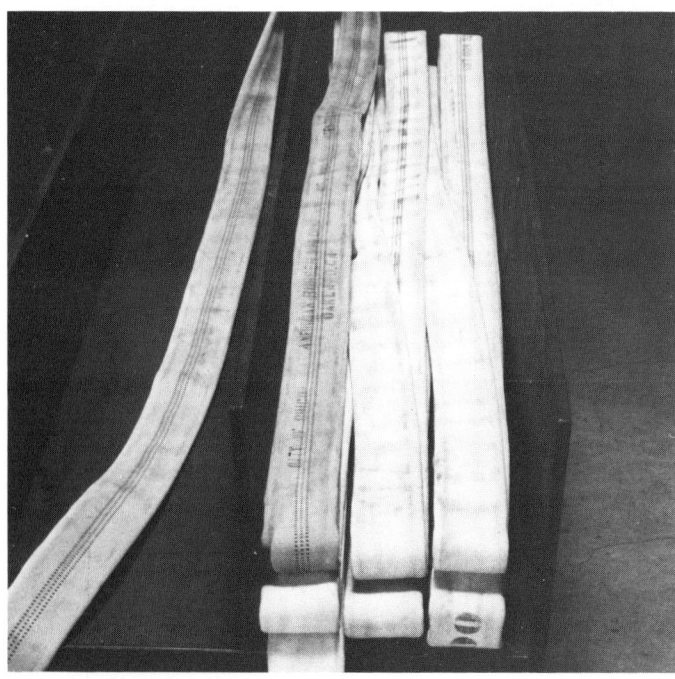

Figure 5.32 STEP 5: Make the folds of the second tier approximately 2 inches (51 mm) shorter than the folds of the first tier.

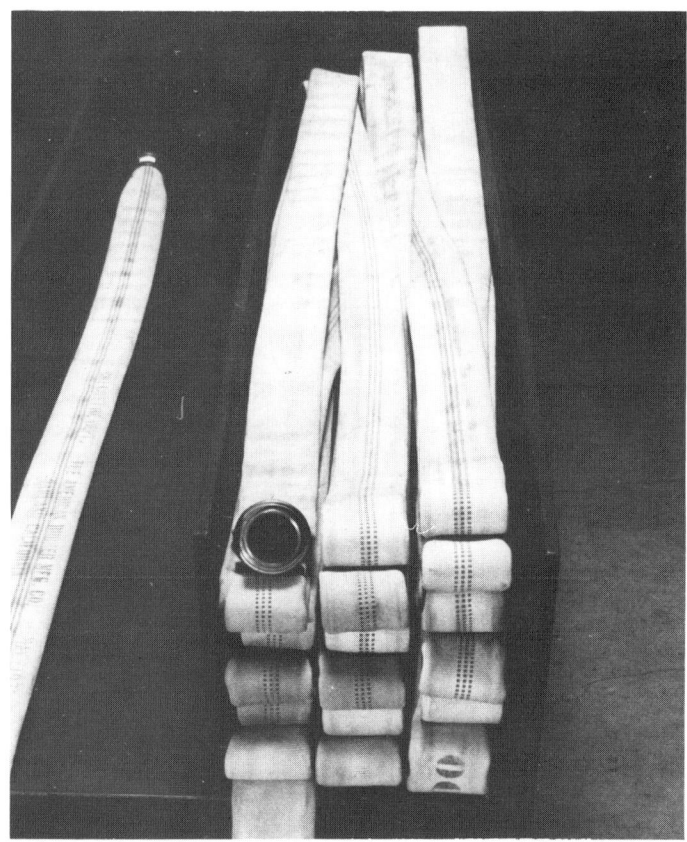

Figure 5.34 STEP 7: Start the load on the opposite side by placing the male coupling against the front wall of the hose bed. Load this side in the same manner as the first.

Supply Hose Loads and Layout Procedures **157**

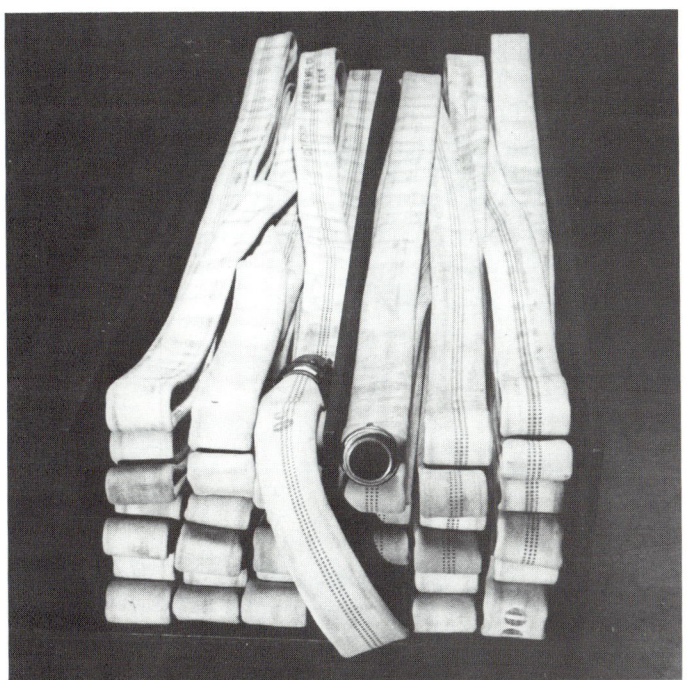

Figure 5.35 STEP 8: Connect the two loads and lay the couplings on top of the load. Pull out the slack so the crossover loop lies tightly against the hose load.

This method can be adapted for loading LDH. LDH can be loaded directly from the street after an incident by straddling the hose with the pumper and driving slowly forward as the hose is progressively loaded into the bed. A hose wringer or roller should be used to expel the air and water from the hose as it is placed in the hose bed (Figure 5.36).

THE REEL LOAD

Another way to load hose, especially for LDH, is to put it on a large reel. An advantage with the reel load is that there are no folds in the hose. For this reason, the hose does not need to be reloaded periodically to relieve stress on bends. A disadvantage is that only one line can be laid at a time unless multiple reels are mounted on the apparatus body. Another disadvantage is that travel vibration causes the hose to work loose and hang loosely under the reel. This then causes the reel to hang up when the hose is laid.

An apparatus set up to carry LDH on a reel is often designated as a "wagon" because its primary mission is to lay supply hose for other apparatus. The reel is usually mounted on the rear of the apparatus and should be hydraulically powered to aid in loading (Figure 5.37).

Figure 5.36 Use a hose wringer to expel air and water from LDH hose before reloading.

Figure 5.37 An LDH hose reel is usually mounted at the rear of the apparatus.

When reel-loading LDH in the field, the simplest method is the same as for loading the flat load. Straddle the hose with the apparatus and progressively load it as the vehicle moves forward slowly. Another method used in the station for loading LDH hose from a storage roll involves the use of a hose loading table, the top of which rotates (Figure 5.38). The roll is simply placed on the table and unrolled as it is loaded onto the reel (or the hose bed, as the case may be).

The procedure for loading LDH onto a reel is as follows:

Step 1: Place the first coupling on the reel to one side (Figure 5.39).

Step 2: Rotate the reel to start the load. Make the first wrap over the coupling to anchor it to the reel (Figure 5.40).

Figure 5.38 The top of a hose loading table rotates to lay out hose from a hose roll as it is loaded into the hose bed. *Courtesy of the Chico, California Fire Department.*

Figure 5.39 STEP 1: Place the first coupling on one side of the reel. *Courtesy of Lower Providence Township, Pennsylvania Volunteer Fire Company.*

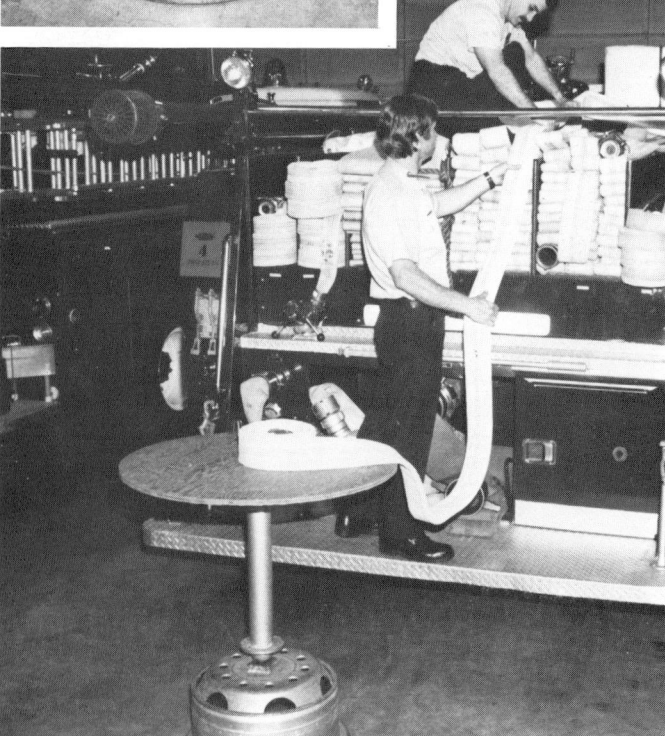

Figure 5.40 STEP 2: Make the first wrap over the coupling to anchor it to the reel. *Courtesy of Lower Providence Township, Pennsylvania Volunteer Fire Company.*

Step 3: Using the power-wind switch, rotate the reel to layer the hose on the reel. Angle the hose on the reel so that each wrap is laid progressively across the reel (Figure 5.41).

Step 4: Continue to load the hose, progressively building each layer in the same manner as the first (Figure 5.42). Avoid laying couplings directly over previously loaded couplings.

Step 5: When the hose is completely loaded, set the hose reel brake to secure the reel in a stationary position.

Figure 5.41 STEP 3: Use the power-wind switch to rotate the reel and load the hose on the reel. Angle the hose on the reel so that each wrap is laid progressively across the reel. *Courtesy of Lower Providence Township, Pennsylvania Volunteer Fire Company.*

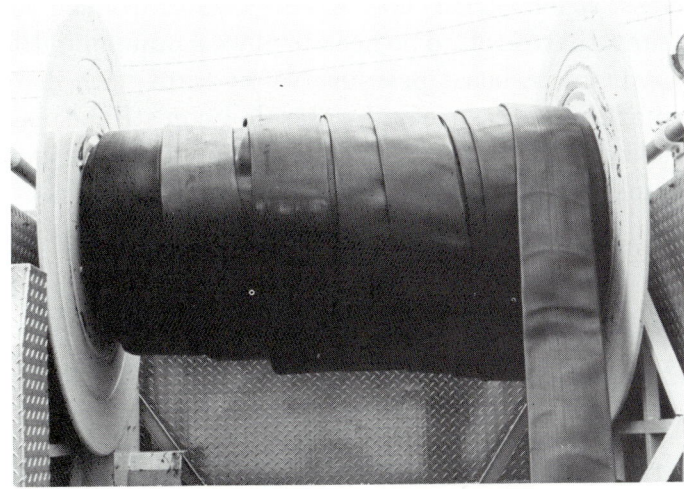

Figure 5.42 STEP 4: Continue to load the hose in the same manner to add layers of hose to the reel. *Courtesy of Lower Providence Township, Pennsylvania Volunteer Fire Company.*

HOSE LAY PROCEDURES

Layout procedures often vary from department to department, but the basic methods of laying hose remain the same. Hose is either laid *forward* from a water source to the incident scene, *reverse* from the incident scene to a water source, or *split* so that the hose can be laid to and from the junction to the water source and the incident scene. These basic methods are presented to provide the foundation for developing hose lays that more specifically suit individual department needs.

When laying hose, a few basic guidelines should be followed:

- Personnel should avoid riding the tailboard when the hose is being laid from a moving apparatus (Figure 5.43).

- Drive at a speed no greater than that which allows the couplings to clear the tailboard

Figure 5.43 Do NOT ride the tailboard while laying out hose from the hose bed.

as the hose leaves the bed — generally between 5 and 10 mph (8 km/h and 16 km/h) (Figure 5.44).

- Lay the hose to one side of the roadway so that other apparatus are not forced to drive over it (Figure 5.45).

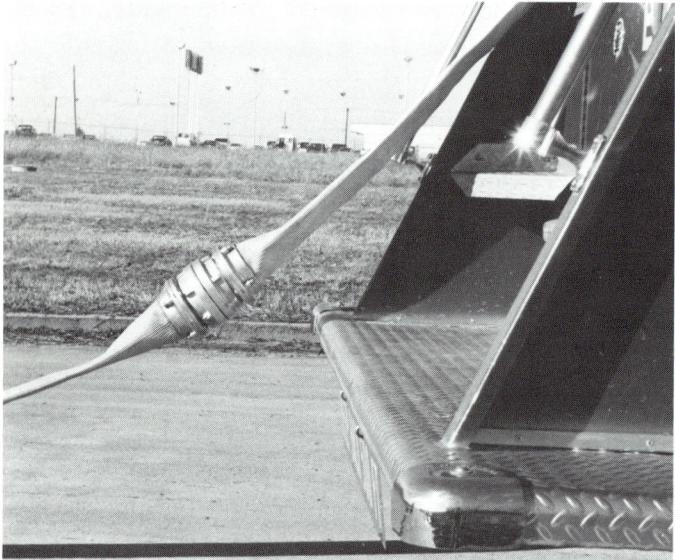

Figure 5.44 Drive the apparatus between 5 mph and 10 mph (8 km/h and 16 km/h) so that the couplings clear the tailboard as the hose comes from the hose bed.

Figure 5.45 Lay the hose to one side of the roadway so other apparatus are not forced to drive over it.

Determining Which Hydrant to Use

In jurisdictions where the primary water source is a water system with hydrants, it is usually best to use the hydrant closest to the fire. This minimizes the length of hose laid and thus minimizes the friction loss in the hose. As mentioned earlier, some officers prefer to go directly to the incident scene to make a size-up before committing the apparatus to laying a supply hose. In a department where this is a standard procedure, apparatus are usually set up for reverse lays. In this case, it is not unusual for the pumper to lay to the closest hydrant *beyond* the fire.

There are situations when the closest hydrant may not be the best choice. In some cases, it may be necessary to select a more distant hydrant. For example, if the closest hydrant has a significantly smaller flow than a more distant hydrant (perhaps because the closest hydrant is on a smaller water main than the distant hydrant), it may be better to lay to the hydrant with the greater flow. A more distant hydrant may also be better if the closest hydrant is located too near an intensely burning building. Use of the closest hydrant would, in this case, unnecessarily expose the apparatus and personnel to heat from the fire. It may also be better to choose a distant hydrant on the same side of the street as the fire building rather than a closer hydrant on the opposite side of the street. This is especially important when LDH is laid because, if laid across a street, it impedes passage of incoming fire apparatus.

In spite of the potential traffic problem with LDH, its use makes hydrant selection much easier. An LDH-equipped pumper can lay hose in either direction because the couplings are usually sexless. When LDH is used to carry water from a hydrant, there is often no need to place a pumper at the hydrant to boost pressure. A hydrant with even marginal pressure and water flow can still be used because friction loss in LDH carrying a lesser volume of water is virtually nonexistent.

Making a Hydrant

When a pumper lays hose forward from a hydrant, the hose should be anchored to the hydrant so that as the pumper proceeds to the incident scene, the hose is pulled from the bed and laid in the street. Anchoring the hose in this manner is safer than having a person stand in the street holding the end of the hose. If a coupling hangs up in the hose bed, or if the hose becomes entangled, the hydrant will better withstand the sudden jerk of the hose than if it is held by a firefighter in a free-standing position. Anchoring the hose to the hydrant also allows the pumper to lay hose without the necessity of leaving someone at the hydrant. In this case, the hose is laid and left uncharged until a second pumper and crew arrive to make the connection (or the traditional phrase, "make the hydrant").

Communication is important when a pumper stops at a hydrant to lay hose in a forward lay. The firefighter assigned to pulling hose and anchoring it to the hydrant should wait for a direct order before doing so. The firefighter should not act automatically whenever the pumper stops at a hydrant. This is because the officer in charge may have the pumper pause at a hydrant to make an assessment of the fire situation from that vantage point. If the officer chooses to use another hydrant, premature action by the firefighter could have a detrimental effect on the fire fighting operation.

It is also important that the officer communicate whether the supply hose should be charged as soon as the firefighter makes the connection. An option here is for the firefighter's action to be predetermined by policy. Some departments specify that whenever a forward lay is made, the hydrant should be charged as soon as possible. This, of course, requires that someone in the crew take immediate action when the pumper stops at the incident scene. That person is responsible for either clamping the hose near the pumper or disconnecting the hose from the hose bed and immediately connecting it to an intake.

Another option for charging a supply line is to have the firefighter wait for a verbal or visual signal before opening the hydrant. This can be done by radio (if the firefighter carries a portable unit), by sounding the air horn on the pumper (for example, three short blasts can mean "open the hydrant"), or by a hand signal (for example, two upstretched arms can mean "charge the line") (Figure 5.46).

The procedure for anchoring and connecting hose to a hydrant starts when the driver stops the apparatus with the tailboard approximately 10 feet (3 m) past the hydrant. The officer then gives

Figure 5.46 Two upstretched arms is the signal for "charge the line."

the appropriate command (for example, "Make the hydrant" or "Forward lay, one line") for the firefighter to make the hydrant. The procedure for making a hydrant is as follows (a four-way valve is preattached to the hose):

Step 1: Pull enough hose to wrap around the hydrant (Figure 5.47) (take along the hydrant tools).

Figure 5.47 STEP 1: Pull enough hose to wrap around the hydrant.

Step 2: Wrap the hydrant with the hose and signal the driver to lay the hose (Figure 5.48).

Step 3: After the pumper has laid several sections of hose, unwrap the hose from the hydrant and attach the four-way valve and hose to the appropriate outlet (Figure 5.49).

Step 4: Set the four-way valve to route water into the hose, then open the hydrant valve at the appropriate time (Figure 5.50).

Figure 5.50 STEP 4: Set the four-way valve to the proper position, then open the hydrant at the appropriate time.

Figure 5.48 STEP 2: Wrap the hose around the hydrant and signal the driver to proceed to the fire.

Figure 5.49 STEP 3: Unwrap the hose and connect the four-way valve and hose to the outlet.

The Forward Lay

When laying hose forward from the water source to the incident scene, the pumper may be used in several ways. The option chosen often depends upon the size of hose laid. It is important to remember that supply hose delivers a volume of water directly proportional to its size. A rule of thumb is given in Table 5.1.

It follows, therefore, that one option when LDH is laid is to simply spot the pumper at the most tactically advantageous place at the incident

TABLE 5.1 AMOUNT OF WATER DELIVERED BY SUPPLY HOSE	
INCHES (mm)	**GPM (L/min)**
2½-inch hose (65 mm hose)	200 - 250 gpm (757 L/min - 946 L/min)
3-inch hose (77 mm hose)	300 - 350 gpm (1 136 L/min - 1 325 L/min)
4-inch hose (100 mm hose)	750 - 1,000 gpm (2 839 L/min - 3 785 L/min)
5-inch hose (125 mm hose)	1,200 - 1,500 gpm (4 542 L/min - 5 678 L/min)

scene, connect the LDH to an intake valve, and pump the water from this supply hose.

If medium diameter hose is laid and the crew is involved in a quick-attack operation using preconnected attack lines, the pump operator may wish to first pump water from the booster tank, then switch to the supply hose when it becomes charged with water. The critical point in this situation is that fire fighting crews could be placed in a dangerous situation if the booster tank water becomes depleted before the supply line is charged. The option of starting a quick attack using water from the booster tank should be chosen only if the crew is reasonably certain it can achieve knockdown using only the water in the tank or if the supply hose can be connected and charged within one or two minutes.

Another option when forward laying medium diameter hose is for the pumper to drop fire fighting tools and hose at the incident scene and return to the hydrant to boost pressure. When this happens, the forward-laid hose becomes an attack hose. (The procedure for an attack-line forward lay is described in Chapter 6.)

This procedure for making a forward lay (Figure 5.51) can be modified to accommodate the apparatus, hose, and equipment used:

Step 1: Stop the apparatus at the hydrant, remove the appropriate hydrant tools, and pull sufficient hose to make the hydrant.

Step 2: When signaled by the hydrant firefighter, drive forward to lay out the hose.

Step 3: Stop the apparatus at the appropriate place. Apply a hose clamp and signal for the line to be charged (hydrant firefighter will have made the hydrant).

Step 4: Pull the remaining length of the section of hose coming from the hose bed and disconnect the couplings (return the female coupling to the hose bed).

Step 5: Connect the hose to the intake valve, release the hose clamp, then open the intake valve.

The LDH Supply Lay

The forward lay procedure for LDH is basically the same as the forward lay, except that the hose is not clamped (LDH should *not* be clamped). The following is a simple procedure for laying LDH forward from the hydrant:

Step 1: Stop the apparatus at the hydrant to allow the firefighter to remove the appropriate hydrant tools (including a hydrant adap-

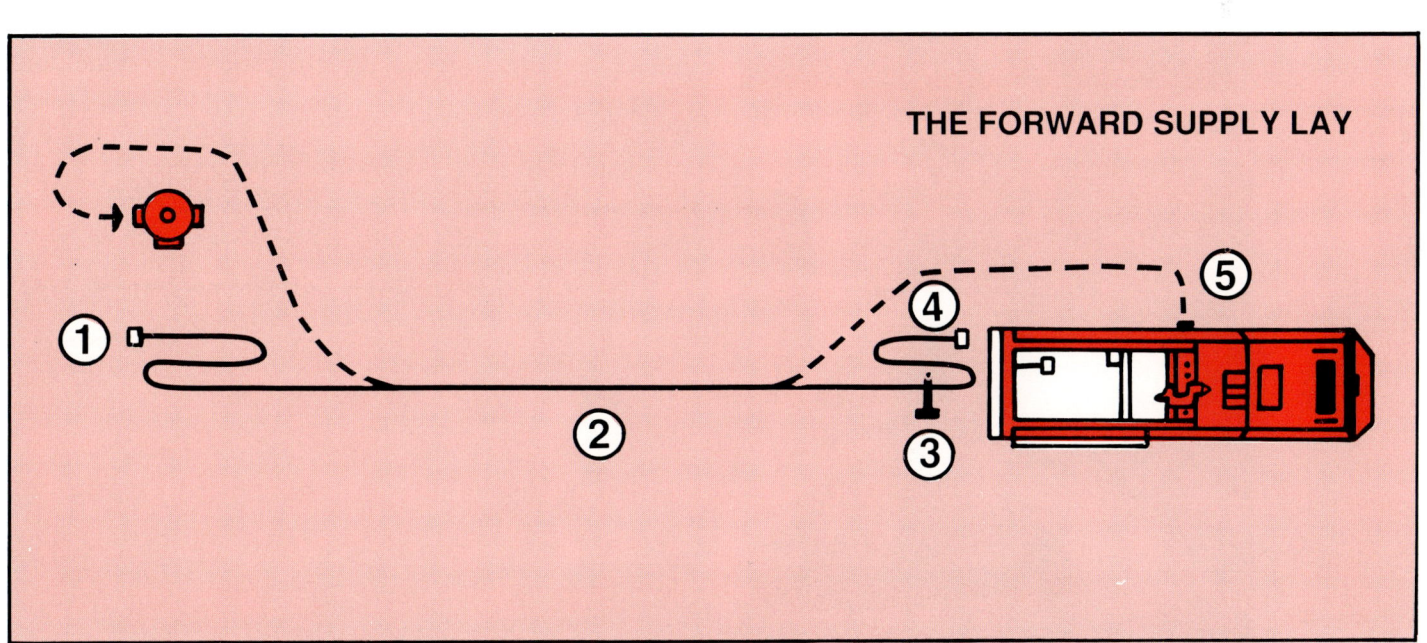

Figure 5.51 The Forward Supply Lay. STEP 1: Pull sufficient hose to make the hydrant. STEP 2: Lay out the hose. STEP 3: Apply a hose clamp and signal for the line to be charged (hose connected to hydrant). STEP 4: Disconnect the last section of hose coming from the hose bed. STEP 5: Connect the hose to the intake valve, release the hose clamp, then open the intake valve.

ter) and pull sufficient hose to make the hydrant (Figure 5.52).

Step 2: When signaled by the hydrant firefighter, drive forward to lay out the hose (Figure 5.53).

Step 3: Stop at the appropriate place. Pull the remaining section of hose from the hose bed, break the connection, and reconnect the LDH coupling to the intake valve (Figure 5.54).

Figure 5.52 STEP 1: Stop the apparatus at the hydrant to allow the firefighter to remove the appropriate hydrant tools and pull sufficient hose to make the hydrant.

Figure 5.53 STEP 2: When signaled by the hydrant firefighter, drive forward to lay out the hose.

Step 4: Signal for water (Figure 5.55) and open the intake valve.

Figure 5.54 STEP 3: Break the connection and reconnect the LDH coupling to the intake valve.

Figure 5.55 STEP 4: Signal for water and open the intake valve.

The Reverse Lay

Laying hose from the incident scene back to the water source has become a standard method for setting up a relay pumping operation when using medium diameter hose as a supply line. In most cases with medium diameter hose, it is necessary to place a pumper at the hydrant to supplement hydrant pressure to the supply hose. It is, of course, always necessary to place a pumper at the water source when drafting. The reverse lay is the most direct way to accomplish this.

A common operation involving two pumpers — an attack pumper and a water supply pumper — calls for the first-arriving pumper to go directly to the scene to start an initial attack on the fire, while the second-arriving pumper lays the supply line from the attack pumper back to the water source. This is a relatively simple operation because the second pumper needs only to connect its just-laid hose to a discharge, connect a suction hose, and begin pumping.

When reverse-laying a supply hose, it is not necessary to use a four-way hydrant valve. One can be used, however, if it is expected that the pumper will later disconnect from the supply hose and leave the hose connected to the hydrant. This may be desirable when the demand for water diminishes to the point that the second pumper can be made available for response to other incidents. As with a forward lay, using the four-way valve in a reverse lay provides the means to switch from pump pressure to hydrant pressure without interrupting the flow.

The reverse lay is also used when the first-in pumper arrives at a fire and must work alone for an extended period of time. In this case, the hose laid in reverse becomes an attack line. It is often connected to a reducing wye so that two smaller hoses can be used to make a two-directional attack on the fire (the attack-line reverse lay is described in greater detail in Chapter 6).

The reverse lay procedure described is for the second-in pumper to lay line from an attack pumper to a hydrant (Figure 5.56 on next page). It can be modified to accommodate most types of apparatus, hose, and equipment.

166 HOSE

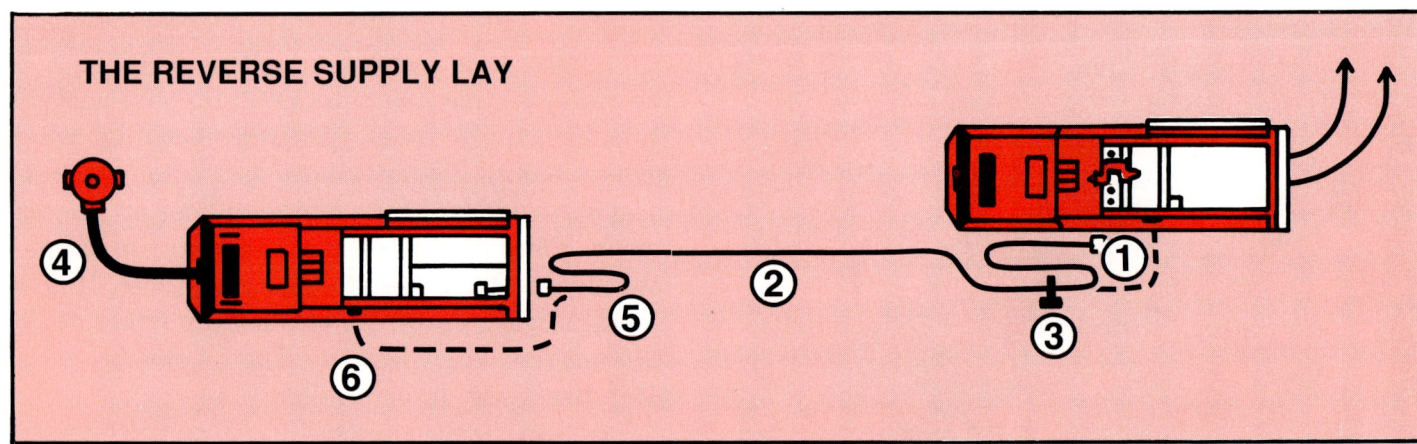

Figure 5.56 The Reverse Lay. STEP 1: Pull sufficient hose to reach the intake valve. STEP 2: Lay out the hose to the water source. STEP 3: Apply a hose clamp to the hose. STEP 4: Make a suction hose connection. STEP 5: Disconnect the last section of hose coming from the hose bed. STEP 6: Connect the supply hose to a discharge valve and charge the hose.

Step 1: Stop the apparatus so that its tailboard is slightly past the intake valve of the attack pumper and have a firefighter pull sufficient hose to reach the intake valve.

Step 2: When signaled by the firefighter anchoring the hose, lay out the hose to the water source.

Step 3: A firefighter applies a hose clamp to the hose.

Step 4: Stop the apparatus at the hydrant and make a suction hose connection.

Step 5: Pull the remaining length of the last section of hose coming from the hose bed, disconnect the couplings, and return the male coupling to the hose bed.

Step 6: Connect the supply hose to a discharge valve and charge the hose.

The Split Lay

The term "split lay" can refer to any one of a number of ways to lay multiple supply hoses. Dividing a hose bed into two or more separate sections provides the most options for laying multiple lines. Depending upon whether the beds are set up for forward or reverse lays, lines can be laid in the following ways (assume for now that the hose is in two beds and is of the same diameter):

- Two lines laid forward
- Two lines laid reverse
- Forward lay followed by a reverse lay
- Reverse lay followed by a forward lay
- Two lines laid forward followed by one or two lines laid reverse
- Two lines laid reverse followed by one or two lines laid forward

Clearly, there are many other split lay options when the hose bed is divided. One of the most versatile arrangements is one in which one section of the hose bed contains LDH and the other sections contain small diameter hose that can be used for either supply or attack. A pumper set up in this manner can lay LDH when the fire situation requires the pumper to lay its own supply line and work alone (laying it in forward so the pumper stays at the incident scene). It can use its small diameter hose as a supply line at fires with less demanding waterflow requirements, as well as for attack lines on large fires. A split hose bed, therefore, gives the officer the greatest number of choices when determining the best way to use limited resources.

Supply Hose Loads and Layout Procedures **167**

Chapter 5 Review
Answers on page 236

TRUE-FALSE: Mark each statement true or false. If false, explain why.

1. Supply loads should be designed to allow for the maximum number of layout options to meet the water supply needed in any kind of situation.
   ☐ T   ☐ F _____
   _____

2. Because of its large capacity, LDH is useful only at large, high-volume fires.
   ☐ T   ☐ F _____
   _____

3. The primary advantage of the reverse lay is that the pumper can remain at the incident scene.
   ☐ T   ☐ F _____
   _____

4. LDH should be loaded with all the couplings to the front of the bed.
   ☐ T   ☐ F _____
   _____

5. Hose should be packed tightly to allow the maximum amount to be carried in a bed.
   ☐ T   ☐ F _____
   _____

6. When making a forward lay, the hose should be anchored to the hydrant as the pumper proceeds toward the fire scene.
   ☐ T   ☐ F _____
   _____

7. On a pumper equipped for a forward lay, the responsible firefighter should automatically begin pulling hose when the pump stops at a hydrant.
   ☐ T   ☐ F _____
   _____

8. LDH should not be clamped.
   ☐ T   ☐ F _____
   _____

168 HOSE

SELECT: Circle the correct response.

When hose is used in a/an **9.** (attack, supply) capacity, it transports water from the source to the pump; when it is used in a/an **10.** (attack, supply) capacity, it transports water from the pump to the nozzle(s).

An apparatus with a large-capacity pump should carry a hose load sufficient to deliver a volume of water equal to the **11.** (average, maximum) output of the pump.

Threaded-coupling hose must be arranged in the hose bed so that when hose is laid, the **12.** (male, female) coupling is toward the water source and the **13.** (male, female) coupling is toward the fire.

FILL IN THE BLANK: Fill in the blanks with the correct values.

14. NFPA 1901 recommends that a minimum of _____ feet (_____ m) of 2½-inch (65 mm) or larger hose be carried on an apparatus.

15. A rule of thumb regarding the volume of water delivered by each size of supply hose is
    A. 2½-inch (65 mm) hose delivers _____ gpm (_____ L/min)
    B. 3-inch (77 mm) hose delivers _____ gpm (_____ L/min)
    C. 4-inch (100 mm) hose delivers _____ gpm (_____ L/min)
    D. 5-inch (125 mm) hose delivers _____ gpm (_____ L/min)

SHORT ANSWER: Answer each item briefly.

16. How many 2½-inch (65 mm) hoses are required to deliver the same amount of water as a single 5-inch (125 mm) hose, assuming the pump pressure is the same for both?

17. How far will the pump pressure required to move 1,000 gpm (3 785 L/min) through two 100-foot (30 m) 2½-inch (65 mm) hoses move the same amount of water through a single 5-inch (125 mm) hose?

18. Which type of hose lay is used when hose is laid from the water source to the fire?

19. In the lay referred to in question 21, which coupling on threaded hose should come off the hose bed first?

20. When loading the bed for the type of lay referred to in question 21, which end of the coupling should be placed in the hose bed first?

21. Which type of hose lay is used when hose is laid from the fire to the water source?

22. In the lay referred to in question 24, which coupling on threaded hose should come off the hose bed first?

23. When loading the bed for the type of lay referred to in question 24, which end of the coupling should be placed in the hose bed first?

24. Which type of hose lay should be used when water is obtained by drafting — forward or reverse?

25. In which type of lay is a hoseline laid in both directions?

26. Which hose loads are recommended for LDH?

27. What is the recommended procedure for loading LDH in a flat load after a fire?

28. Which type of hose lay is typically used by officers who wish to make a size-up of the incident scene before committing the apparatus to laying a supply hose?

170 HOSE

LISTING

29. List the four factors that should be considered when designing supply hose loads.

   A. _____
   B. _____
   C. _____
   D. _____

30. List the three basic hose lays for supply hose.

   A. _____
   B. _____
   C. _____

31. List the three most basic types of hose loads for a supply hose bed.

   A. _____
   B. _____
   C. _____

32. List in order the steps to be followed when making a hydrant.

   Step 1: _____
   Step 2: _____
   Step 3: _____
   Step 4: _____
   Step 5: _____
   Step 6: _____
   Step 7: _____

33. List the two circumstances when a quick attack using water from a booster line can be started.

   A. _____
   B. _____

DISCUSSION QUESTIONS

Are there any changes you would recommend regarding the type of supply hose load presently used for your apparatus? If so, what and why do you feel these changes would be improvements?

_____
_____
_____
_____
_____
_____
_____

Name some advantages and disadvantages of a reel load for both booster hose and LDH.

_____
_____
_____
_____

LEARNING ACTIVITY
Write a departmental procedure for hose loads that reflects the policy of laying forward, reverse, or combination lays.

6

Attack Hose Loads and Layout Procedures

HOSE

This chapter provides information that addresses performance objectives described in NFPA 1001, *Fire Fighter Professional Qualifications* (1987), particularly those referenced in the following sections:

Fire Fighter I

3-13 Fire Hose, Nozzles, and Appliances

3-13.1

3-13.2

3-13.3

3-13.4

3-13.7

3-13.12

Fire Fighter II

4-13 Fire Hose, Nozzles, and Appliances.

4-13.2

4-13.5

Chapter 6
Attack Hose Loads and Layout Procedures

When hose is used to move water from the pump to nozzles for application on the fire, it is often laid out by hand. As with supply hose, the way attack hose is loaded depends upon a number of factors, including personnel resources, type of apparatus, and size of hose. No matter which arrangement is used, however, it should be loaded so that it can be pulled and deployed in the most expedient manner.

Attack hose can be arranged in four basic ways:

- In an accordion, horseshoe, or flat load (in this way it can be used as either supply or attack hose)
- In a "finish" load attached to the end of a hose bed load
- In a preconnected quick-attack load
- In a portable hose pack

LAYING ATTACK HOSE FROM THE SUPPLY HOSE BED

When a pumper crew must work alone for an extended period of time, as when other apparatus come from distant locations, supply hose must be laid. This often requires the pumper to position at the water source to draft or to boost hydrant pressure. In this solution, the hose laid between the fire and the water source is considered attack hose. The hose can be laid forward or reverse, but in either case, enough must be pulled at the fire scene to reach the most remote part of the structure.

As mentioned before, a disadvantage in placing the pumper at the water source is that fire fighting tools and equipment must be removed from the apparatus prior to its departure to the water source. This causes some delay in the initial attack. A way to expedite this task, however, is to assign specific pieces of equipment to each crew member so that the job is done in the shortest possible time. Another approach is to set up for two different levels of equipment removal. When a complete complement of equipment is needed to support the fire fighting operation for an extended period of time, a "full strip" is made. When a less extensive complement is needed, a "partial strip" is made.

The following equipment, which can be modified to match the apparatus and equipment used, is suggested for a full strip (Figure 6.1): hose, SCBA's and spare bottles, nozzles, forcible entry tools, wye or siamese, axes, ladders, and pike poles.

The following equipment, which can also be modified to match the appartus and equipment

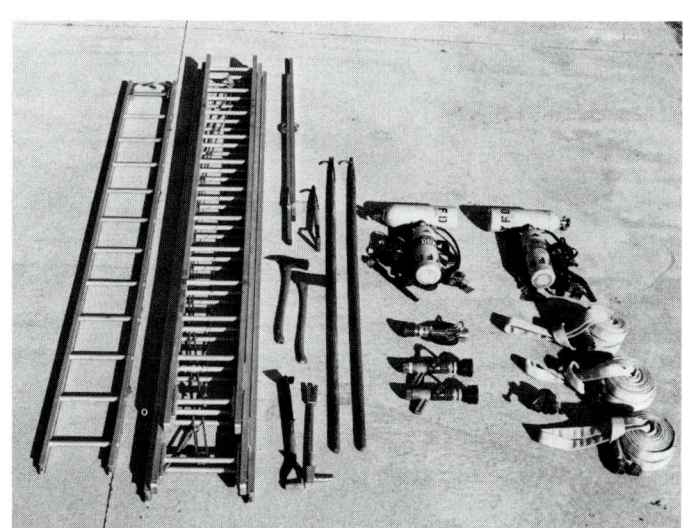

Figure 6.1 A full strip consists of hose, SCBA's and spare bottles, nozzles, forcible entry tools, a wye or siamese, axes, ladders, and pike poles.

Figure 6.2 A partial strip consists of hose, SCBA's, nozzles, forcible entry tools, a wye or siamese, and an axe.

used, is suggested for a partial strip (Figure 6.2): hose, SCBA's, nozzles, forcible entry tools, wye or siamese, and an axe.

Laying a Single Attack Hose in a Forward Lay

When attack hose is laid forward, the hose is laid in the same way as for a forward supply lay, but with one additional step — enough additional hose must be pulled at the scene to reach the rear of the structure. If the hose comes entirely from the supply hose bed, it should be pulled out of the bed in a manner that prevents it from tangling when it is advanced toward the fire.

The following procedure is for forward-laying an attack hose from a flat-loaded hose bed. When sufficient hose and equipment is removed from the pumper, it will return to the hydrant to pump the line from that location. (**NOTE:** Although ordinarily required for this type of evolution, a four-way valve is not shown in Steps 1 and 2.)

Step 1: Stop the apparatus at the hydrant and pull sufficient hose to make the hydrant (Figure 6.3).

Step 2: When signaled by the hydrant firefighter, drive forward to lay out the hose (Figure 6.4).

Step 3: Stop the apparatus at the appropriate place. Apply a hose clamp within 5 feet (1.5 m) of the closest coupling on the hyd-

Figure 6.3 STEP 1: Stop the apparatus at the hydrant and pull sufficient hose to make the hydrant.

Figure 6.4 STEP 2: When signaled by the hydrant firefighter, drive forward to lay out the hose.

Attack Hose Loads and Layout Procedures 177

Figure 6.5 STEP 3: Apply a hose clamp within 5 feet (1.5 m) of the closest coupling on the hydrant side of the coupling.

rant side of the coupling (Figure 6.5). (This may require pulling the remainder of the section coming from the hose bed.)

Step 4: Signal the hydrant firefighter to charge the hose (Figure 6.6).

Step 5: Grasp the two folds (one in each hand) adjacent to the hose that extends from the bed to the street. Pull the hose from the bed so that the folds clear the tailboard by 10 feet (3 m). Lay the folds *on the fire side* of the previously laid hose (Figure 6.7).

Step 6: Pull the next two folds in a similar manner, laying them on the fire side of the previously laid folds (Figure 6.8). Continue in the same manner until sufficient hose is pulled.

Figure 6.7 STEP 5: Pull the first two folds from the bed, clearing the tailboard by 10 feet (3 m), and lay them on the fire side of the previously laid hose.

Figure 6.6 STEP 4: Signal the hydrant firefighter to charge the hose.

Figure 6.8 STEP 6: Lay the next two folds on the fire side of the previously laid folds.

Step 7: Break the connection at the last coupling and return the female coupling to the hose bed (Figure 6.9).

Step 8: Make a full or partial strip of equipment (Figure 6.10).

Step 9: Connect the nozzle and lead in the hose (Figure 6.11).

Step 10: Release the hose clamp (Figure 6.12).

Figure 6.9 STEP 7: Break the connection and return the female coupling to the hose bed.

Figure 6.10 STEP 8: Make a full or partial strip of equipment.

Figure 6.11 STEP 9: Connect the nozzle and lead in the hose.

Figure 6.12 STEP 10: Release the hose clamp safely.

Laying a Single Attack Hose in a Reverse Lay

When attack hose is laid in reverse, the hose is laid in the same way as for laying a reverse supply lay but, again, with one additional step — enough hose must be pulled to reach the rear of the structure. As with a forward lay, the hose should be pulled from the bed in a manner that prevents it from tangling when it is advanced toward the fire.

The following procedure is for reverse-laying hose from a flat-loaded hose bed:

Step 1: Stop the apparatus at the appropriate place. Grasp the male coupling in one hand and the first fold in the other hand

and pull the hose from the bed so that the folds clear the tailboard by 10 feet (3 m). Lay the folds on the ground toward the fire (Figure 6.13).

Step 2: Pull the next two folds in a similar manner, laying them next to the previously laid folds on the side away from the fire (Figure 6.14).

Step 3: Continue in the same manner until sufficient hose is pulled (Figure 6.15).

Step 4: Make a full or partial strip of equipment (Figure 6.16).

Step 5: Anchor the hose and signal the driver to proceed to the hydrant (Figure 6.17).

Step 6: Connect the nozzle and lead the hose in to the point of attack (Figure 6.18).

Figure 6.15 STEP 3: Continue in the same manner until sufficient hose is pulled.

Figure 6.16 STEP 4: Make a full or partial strip of equipment.

Figure 6.13 STEP 1: Pull the male coupling and the first fold from the bed, clearing the tailboard by 10 feet (3 m), and lay the folds on the ground toward the fire. *Courtesy of Ed Sharbar.*

Figure 6.17 STEP 5: Anchor the hose and signal the driver to proceed to the hydrant.

Figure 6.14 STEP 2: Pull the next two folds and lay them next to the previously laid folds on the side away from the fire.

Figure 6.18 STEP 6: Connect the nozzle and lead in the hose.

HOSE LOAD FINISHES

Another way to arrange attack hose is to "finish" a hose load with additional hose that can be quickly pulled at the beginning of a forward or reverse lay. Finishes are arrangements of hose that are usually placed on top of a hose load and are connected to the end of the load.

Finishes fall into two categories: those for forward lays and those for reverse lays. A finish for a reverse load expedites making a full or partial strip of equipment for fire fighting. Finishes for forward lays are usually designed to speed the pulling of hose when making a hydrant and, therefore, are not as elaborate as finishes for reverse lays.

Finish for a Forward Lay

A forward lay finish provides additional hose to reach from the place where the pumper starts the forward lay to the water source. This might be necessary when the pumper cannot easily get to a hydrant such as when the hydrant is set back on the narrow drive within an apartment complex. When a finish is used in a forward lay, the first step for laying the hose is to pull the finish from the hose bed. From this point on, the lay progresses as usual with the firefighter anchoring the hose by holding it near the point where the finish is connected to the hose load.

One of the simplest finishes is a donut roll attached to the end of the supply load. In this case, however, roll the donut so that the male coupling is on the outside of the roll. Then place the donut roll on top of the hose load and connect the male coupling to the female coupling at the free end of the load (Figure 6.19). To pull the donut roll from the hose bed, simply grasp the female coupling and give a sharp tug to flip the roll from the hose bed to the street (Figure 6.20). After the apparatus has laid the hose down the street, the firefighter anchoring the hose grasps the female coupling and hydrant tools and leads the hose in to the hydrant to make the connection (Figure 6.21).

Figure 6.19 Connect the donut roll to the free end of the load.

Figure 6.20 With a sharp tug, flip the roll from the hose bed to the street.

Figure 6.21 Pick up the hydrant tools and lead the hose in to the hydrant to make the connection.

Finishes for Reverse Lays

The primary advantage of a reverse load hose finish is that the entire length of hose needed to reach the structure from the street can be pulled in one motion. Most of the finishes described here use a gated wye, which provides better control of incoming water than a hose clamp. A disadvantage, however, is that if the hose load to which the finish is connected must be laid as a supply line, the finish must be disconnected and removed from the load to permit laying the underlying hose.

Three finishes are described here: the cisco, the reverse horseshoe, and the skid load. Each is designed so that one person can pull the entire finish from the bed.

CISCO FINISH

The cisco finish is made of two 100-foot (30 m) lengths of hose, each connected to one side of a wye. Any size of attack hose can be used: 1½-, 1¾-, or 2½-inch (38 mm, 45 mm, or 65 mm). The smaller sizes require a 2½ x 1½-inch (65 mm by 38 mm) gated reducing wye; the 2½-inch (65 mm) hose requires a 2½ x 2½-inch (65 mm by 65 mm) gated wye. Two nozzles of the appropriate sizes are also needed. The finished bundle, which is accordion folded on top of the hose load, is tied together with straps or ropes.

This procedure is for making a cisco finish with 1½-inch (38 mm) hose:

Step 1: Connect the wye to the end (male) coupling of the hose load at the rear of the bed, then place it in the center of the hose load with the two male openings toward the rear of the bed (Figure 6.22).

Step 2: Connect one of the 1½-inch (38 mm) hoses to one side of the wye and lay the first fold of hose on edge to the front. Make a fold at the front and lay the hose on edge back to the rear of the bed (Figure 6.23).

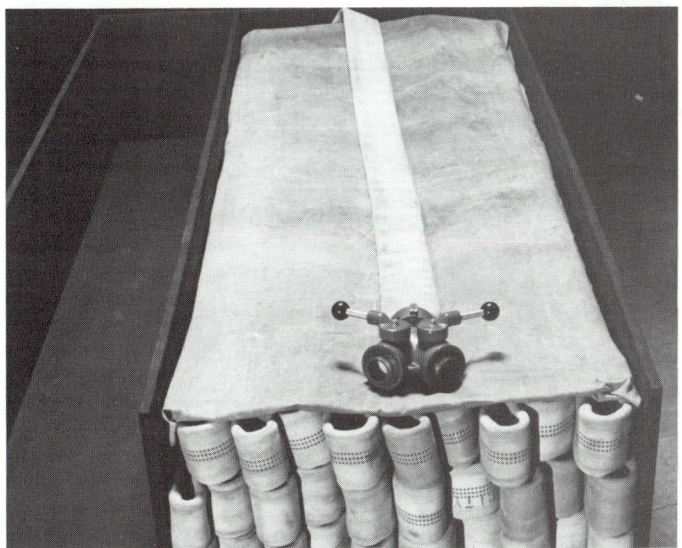

Figure 6.22 STEP 1: Connect the wye to the hose load, then place it in the center of the hose load.

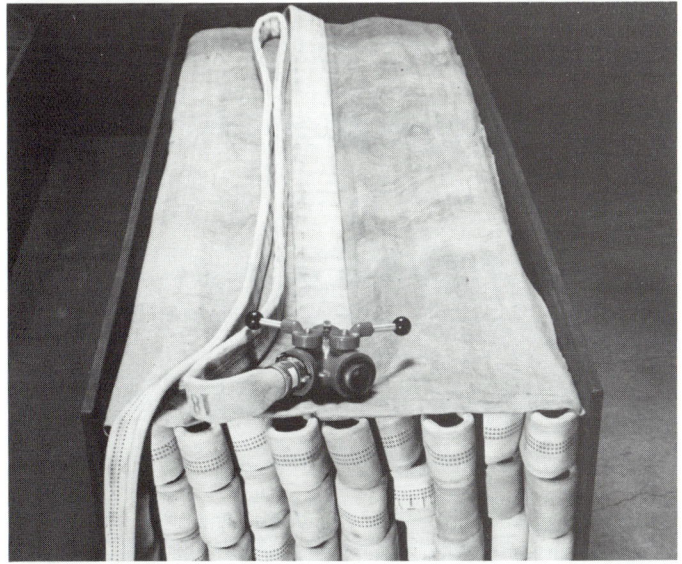

Figure 6.23 STEP 2: Connect one of the 1½-inch (38 mm) hoses to the wye and lay the first fold of hose on edge to the front. Make a fold at the front and lay the hose on edge back to the rear of the bed.

182 HOSE

Step 3: Make a fold at the rear near the wye connection, then lay the hose back to the front adjacent to the previous fold (Figure 6.24).

Step 4: Continue accordion folding the hose until the entire length is loaded (Figure 6.25).

Step 5: Connect and load the second length of hose in the same manner as the first, laying it progressively toward the opposite side of the bed (Figure 6.26).

Step 6: Attach a nozzle to each of the two hoses and make sure that the nozzles and wye are in the closed position. Lay the nozzles on top of the finish load and tie each bundle at the rear with a rope or strap. The first fold on each side of the wye should not be tied into the bundle (Figure 6.27).

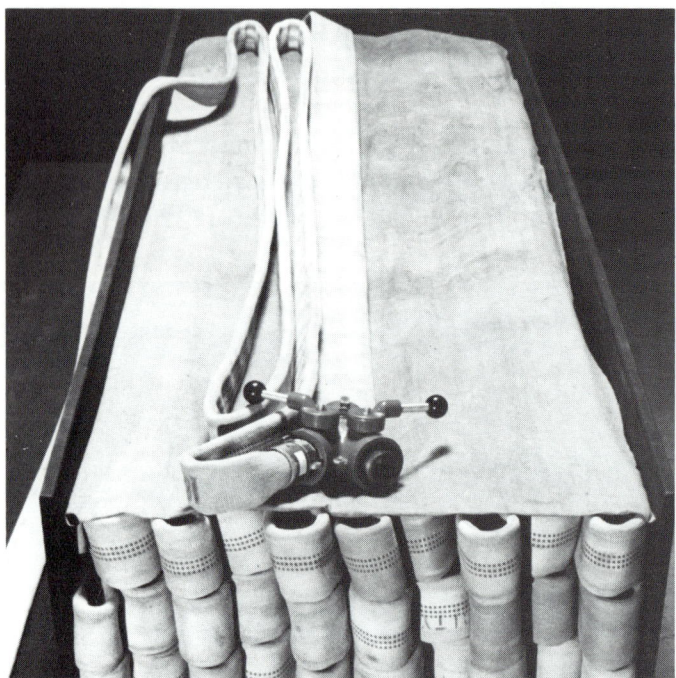

Figure 6.24 STEP 3: Make a fold at the rear near the wye connection, then lay the hose back to the front adjacent to the previous fold.

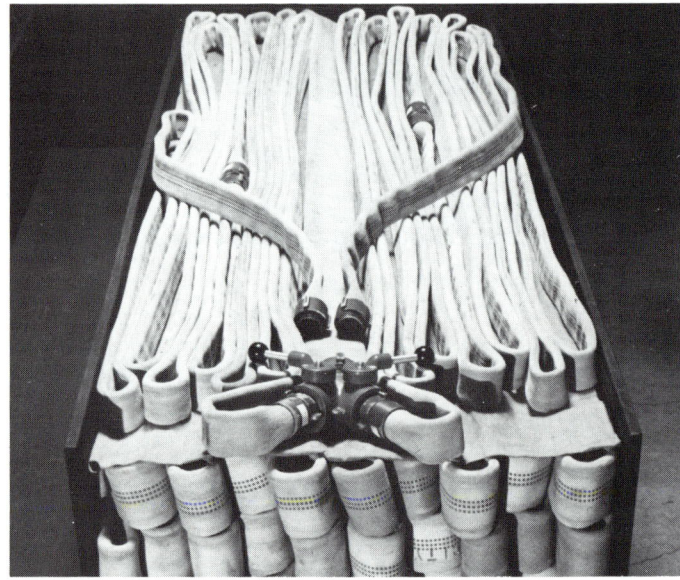

Figure 6.26 STEP 5: Connect and load the second length of hose in the same manner as the first.

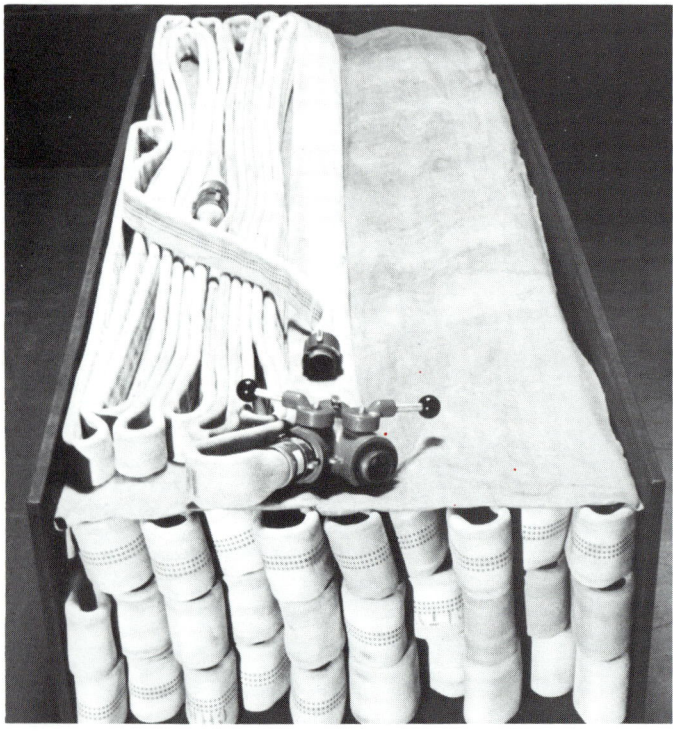

Figure 6.25 STEP 4: Continue accordion folding the hose until the entire length is loaded.

Figure 6.27 STEP 6: Attach the nozzles and lay them on top of the finish load and tie each bundle at the rear with a rope or strap.

Reverse Lay with a Cisco Finish

The cisco finish can be pulled and advanced by one or two persons. The procedure for a reverse lay with a cisco finish is as follows:

Step 1: Grasp each side of one bundle by the tie strap or rope and pull it from the bed. When the end of the bundle clears the tailboard, lay the bundle on the ground (Figure 6.28).

Step 2: Pull the opposite bundle in the same way (Figure 6.29).

Step 3: Pull the wye and attached hose from the bed and lay the wye between the bundles near the ties (Figure 6.30).

Step 4: When you are ready to reverse lay to the hydrant, pick up the wye and signal the driver to proceed. Anchor the hose so that it lays out from the bed as the apparatus proceeds toward the water source (Figure 6.31).

Figure 6.30 STEP 3: Lay the wye between the bundles.

Figure 6.31 STEP 4: Anchor the hose.

Step 5: After the apparatus makes the lay, untie the bundles and lead in the hose (Figure 6.32).

Step 6: Open the wye when ready for water (Figure 6.33).

Figure 6.28 STEP 1: Pull one bundle clear of the tailboard and lay it on the ground.

Figure 6.29 STEP 2: Pull the opposite bundle in the same way.

Figure 6.32 STEP 5: Untie the bundles and lead in the hose.

Figure 6.33 STEP 6: Open the wye when ready for water.

REVERSE HORSESHOE FINISH

This finish is similar to the horseshoe load described in the previous chapter, except that the "U" portion of the horseshoe is at the rear of the hose bed. It is made of two 100-foot (30 m) lengths of hose, each connected to one side of a wye. As with the cisco finish, any size of attack hose can be used 1½-, 1¾-, or 2½-inch (38 mm, 45 mm, or 65 mm). The smaller sizes require a 2½ x 1½-inch (65 mm by 38 mm) gated reducing wye; the 2½-inch (65 mm) hose requires a 2½ x 2½-inch (65 mm by 65 mm) gated wye. Two nozzles of the appropriate size are also needed. The procedure that follows is for making a reverse horseshoe finish with 1½-inch (38 mm) hose:

Step 1: Connect the wye to the end (male) coupling of the hose load at the rear of the bed, then place the wye in the center of the hose load with the two male openings toward the rear of the bed (Figure 6.34).

Step 2: Connect one of the 1½-inch (38 mm) hoses to the wye, then lay the hose on edge to the front of the bed and make a fold (Figure 6.35).

Step 3: Lay the hose back to the rear alongside the first length. Form a "U" at the edge of the bed, then return the hose to the front and make a fold (Figure 6.36).

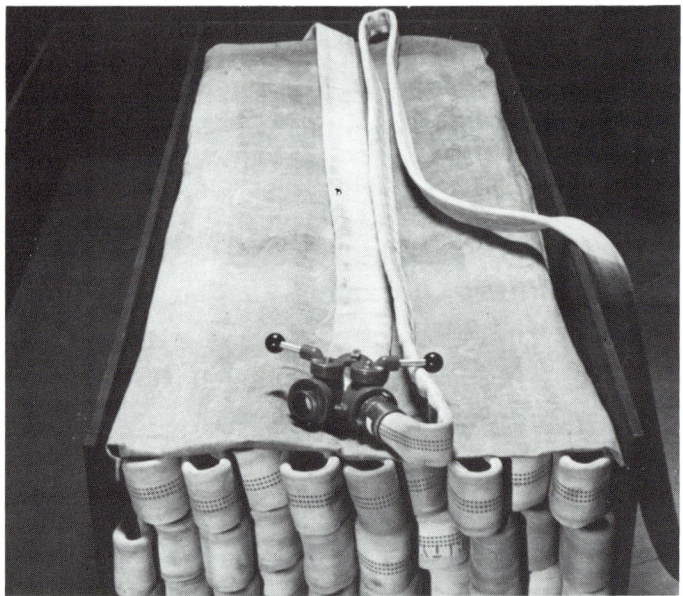

Figure 6.35 STEP 2: Connect one of the 1½-inch (38 mm) hoses to the wye, then lay the hose on edge to the front of the bed and make a fold.

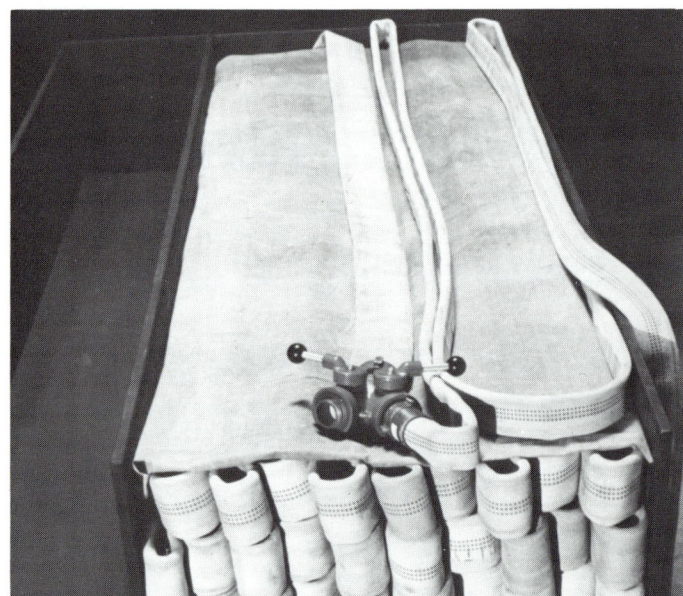

Figure 6.36 STEP 3: Lay the hose back to the rear alongside the first length. Form a "U" at the edge of the bed, then return the hose to the front and make a fold.

Step 4: Lay the hose back inside the previously laid length in the same manner as before and continue until the entire length has been loaded (Figure 6.37).

Step 5: Wrap the male end of the hose once around the horseshoe loops (Figure 6.38).

Step 6: Form a small loop by bringing the end back under the center of the loops, then over the top (Figure 6.39).

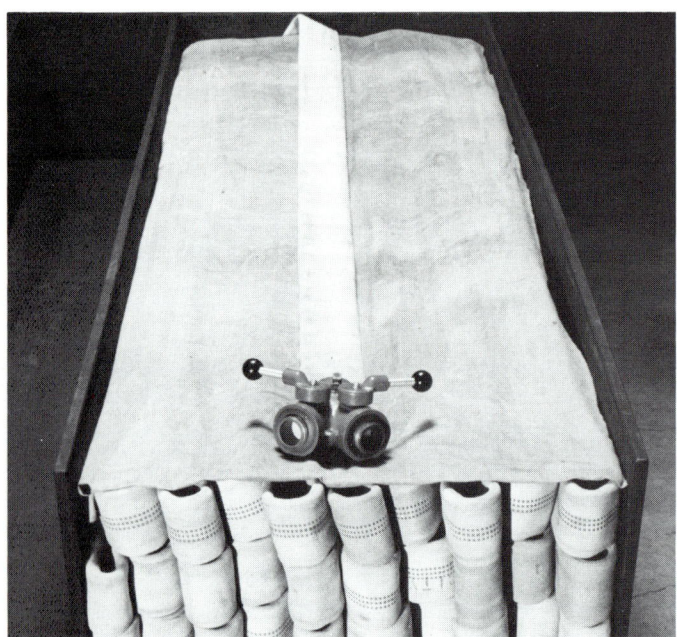

Figure 6.34 STEP 1: Connect the wye and place it in the center of the hose load.

Attack Hose Loads and Layout Procedures **185**

Step 7: Attach the nozzle and place it inside the small loop. Pull the remaining slack hose back into the center of the horseshoe to tighten the loop against the nozzle (Figure 6.40).

Step 8: Load the second length of hose in the same manner on the opposite side of the bed (Figure 6.41).

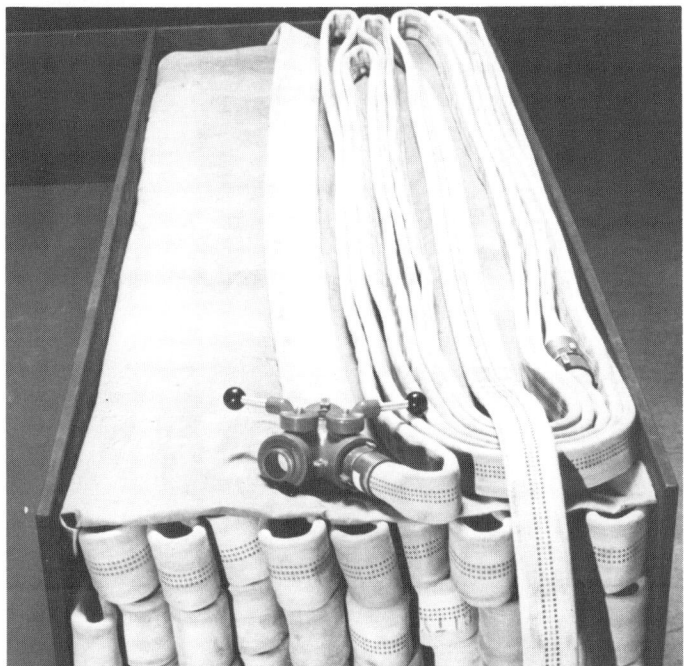

Figure 6.37 STEP 4: Lay the hose back inside the previously laid length in the same manner as before and continue until the entire length has been loaded.

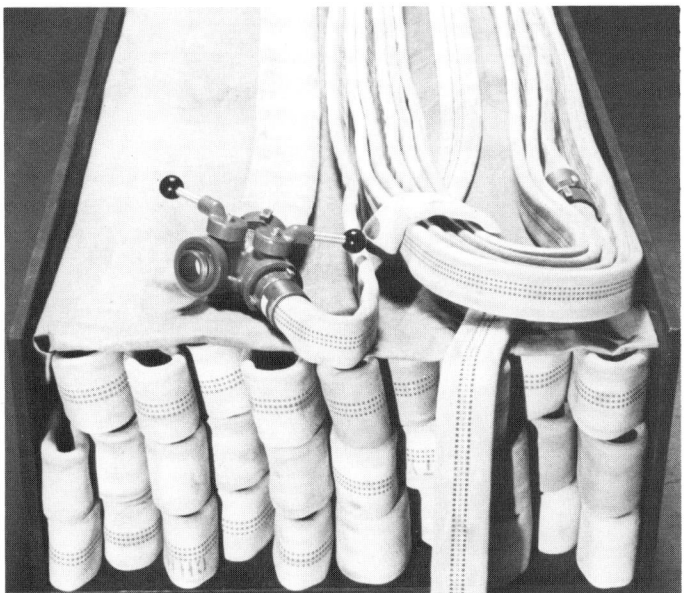

Figure 6.38 STEP 5: Wrap the male end of the hose once around the horseshoe loops.

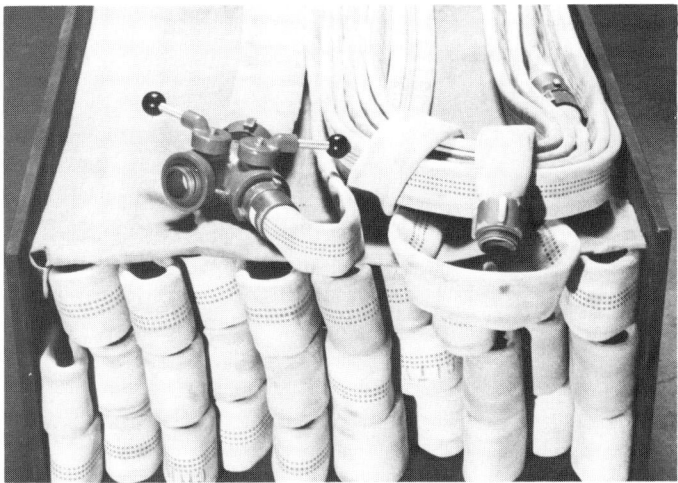

Figure 6.39 STEP 6: Form a small loop by bringing the end back under the center of the loops, then over the top.

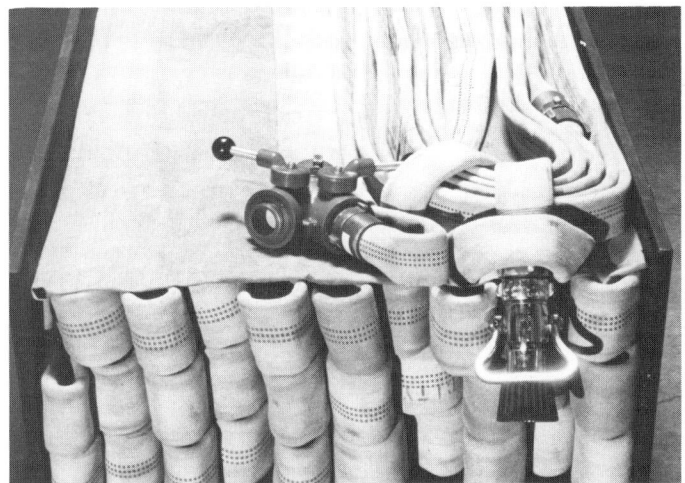

Figure 6.40 STEP 7: Attach the nozzle and place it inside the small loop. Pull the remaining slack hose back into the center of the horseshoe to tighten the loop against the nozzle.

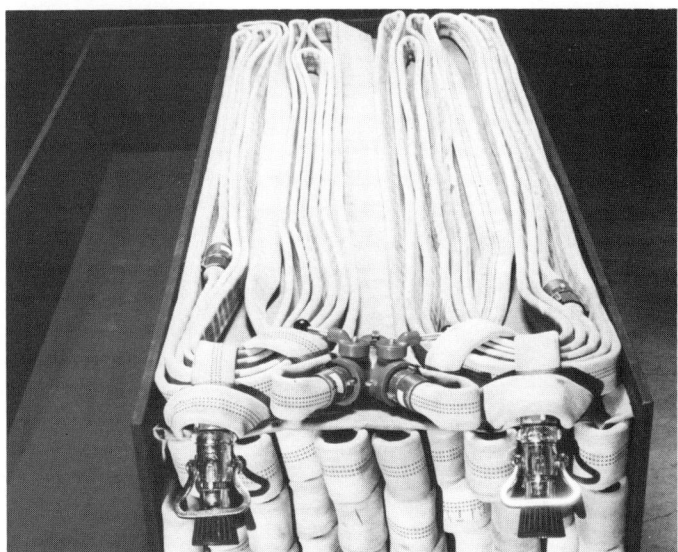

Figure 6.41 STEP 8: Load the second length of hose in the same manner on the opposite side of the bed.

Reverse Lay with a Reverse Horseshoe Finish

The reverse horseshoe finish can also be pulled and advanced by one or two persons. The procedure for a reverse lay with a reverse horseshoe finish is as follows:

Step 1: Grasp the nozzle and small loop of one bundle and pull the bundle from the bed. When the end of the bundle clears the tailboard, lay the bundle on the ground (Figure 6.42).

Step 2: Pull the opposite bundle in the same way (Figure 6.43).

Step 3: Pull the wye and attached hose from the bed and lay the wye between the bundles near the ties (Figure 6.44).

Step 4: When you are ready to reverse lay to the hydrant, pick up the wye and signal the driver to proceed. Anchor the hose so that it lays out from the bed as the apparatus proceeds toward the water source (Figure 6.45).

Step 5: After the apparatus makes the lay, place one arm through the horseshoe loops of one bundle and lay off the loops one at a time to lead in the hose (Figure 6.46). Lay out the second hose in the same manner.

Step 6: Open the wye when ready for water (Figure 6.47).

Figure 6.43 STEP 2: Pull the opposite bundle in the same way.

Figure 6.44 STEP 3: Pull the wye and lay it between the bundles.

Figure 6.42 STEP 1: Pull one bundle clear of the tailboard and lay it on the ground.

Figure 6.45 STEP 4: Anchor the hose.

Attack Hose Loads and Layout Procedures   187

slack hose for pulling the finish from the bed. Turn the hose flat and lay it to the rear of the bed approximately one-third of the way from one side (Figure 6.49).

Figure 6.46 STEP 5: Lay off the loops one at a time as you lead in the hose.

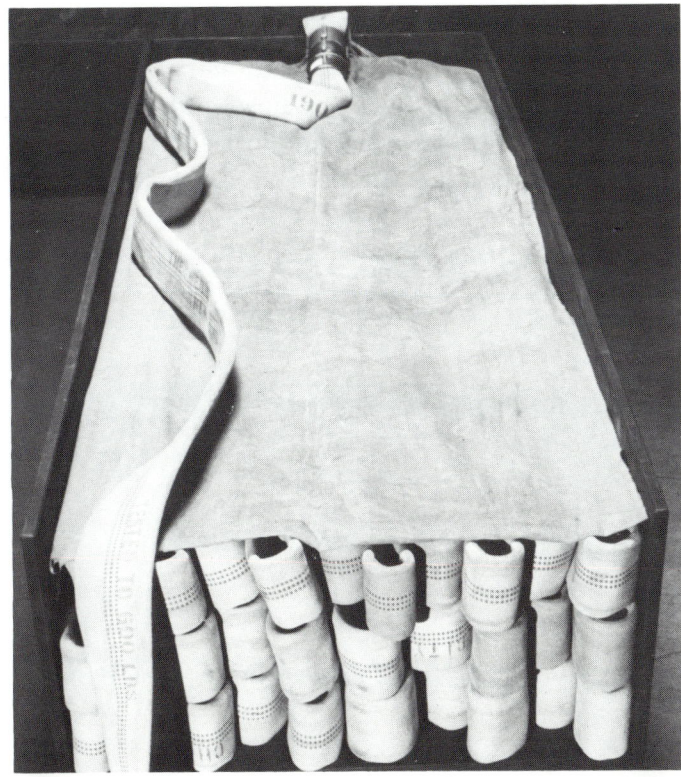

Figure 6.48 STEP 1: Connect the first section of hose to the hose load.

Figure 6.47 STEP 6: Open the wye.

SKID LOAD FINISH

Hose in this finish is loaded in accordion folds across the hose bed. The folds rest on two lengths of hose laid flat to act as skids when the finish is pulled. Three sections of 2½-inch (65 mm) hose and a nozzle are used to make this load. The procedure is as follows:

Step 1: Connect the first section of hose directly to the hose load (Figure 6.48).

Step 2: Lay several folds, accordionlike, on edge across the front of the bed to provide some

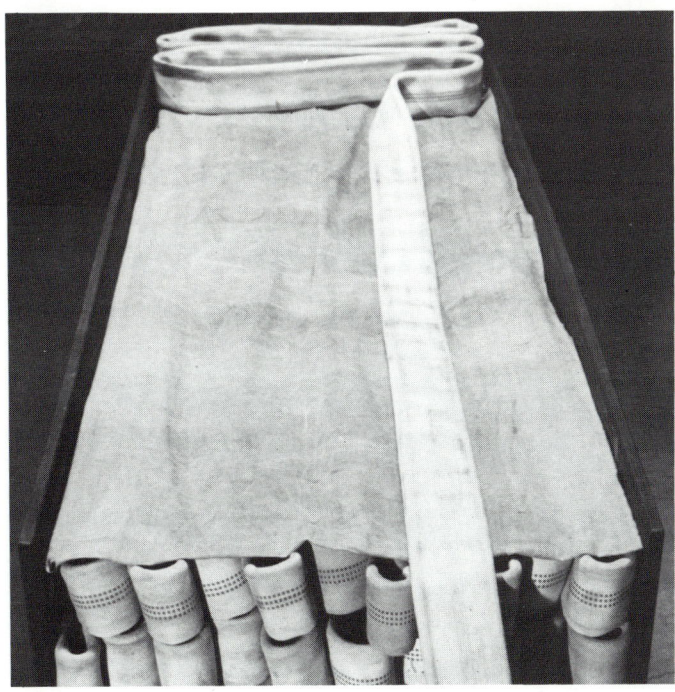

Figure 6.49 STEP 2: Lay several folds, accordionlike, on edge across the front of the bed, then lay it to the rear of the bed approximately one-third of the way from one side.

Step 3: Form a loop that extends approximately 6 inches (152 mm) beyond the hose bed. Lay the hose back to the front of the bed, then over to the other side of the bed approximately one-third of the way from the opposite side. Make the second skid in the same manner as the first (Figure 6.50).

Step 4: Connect additional hose and load it, accordionlike, back and forth across the two skids until all the hose is loaded. Attach the nozzle and place it on top of the load (Figure 6.51).

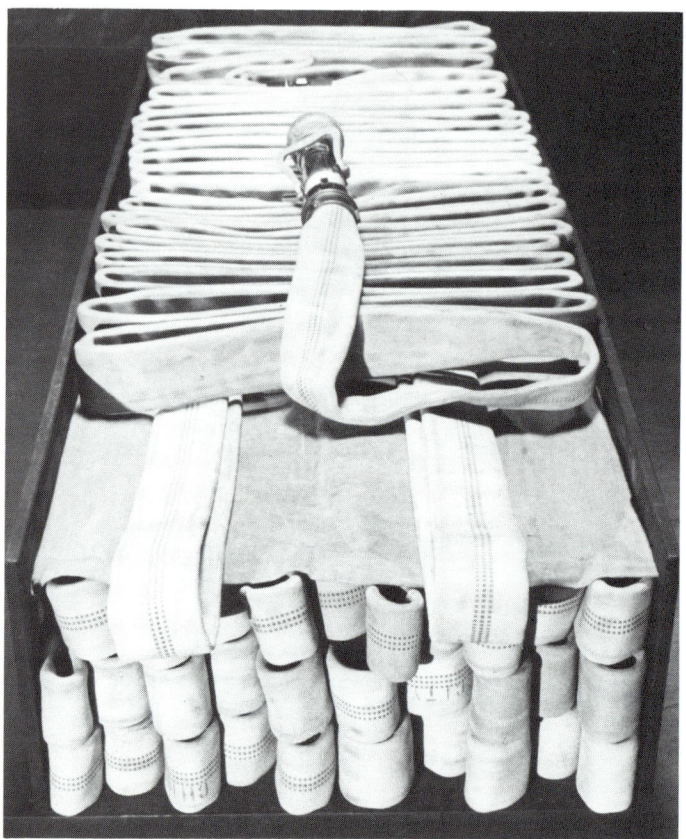

Figure 6.51 STEP 4: Complete the second skid in the same manner as the first, then accordion load additional hose back and forth across the two skids until all the hose is loaded. Attach the nozzle and place it on top of the load.

Reverse Lay with a Skid Load Finish

The skid load finish can be pulled and advanced by one person. The procedure for a reverse lay with a skid load finish is as follows:

Step 1: Grasp the small loops of each skid and pull the finish from the bed. Maintain even tension on both skids to lower the finish from the hose bed to the ground. Pull the finish back from the tailboard approximately 15 feet (5 m) (Figure 6.52).

Step 2: If there is no shutoff valve in the connection between the finish and the hose load, place a hose clamp on the hose approximately 5 feet (1.5 m) from the connection on the apparatus side of the couplings (Figure 6.53).

Step 3: When you are ready to reverse lay to the hydrant, pick up the hose between the finish and the apparatus and signal the driver to proceed. Anchor the hose so that

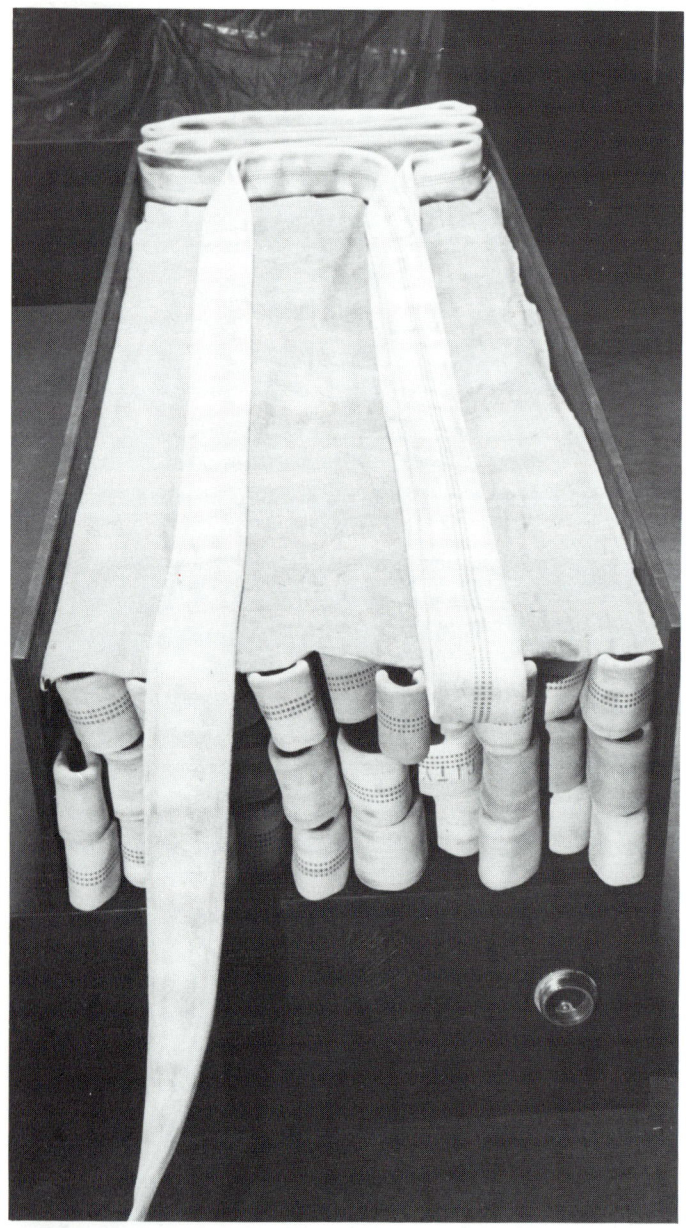

Figure 6.50 STEP 3: Lay the hose flat to the rear of the bed and extend the fold 6 inches (152 mm) beyond the edge. Lay the hose back to the front and around to make the second skid on the opposite side.

Attack Hose Loads and Layout Procedures **189**

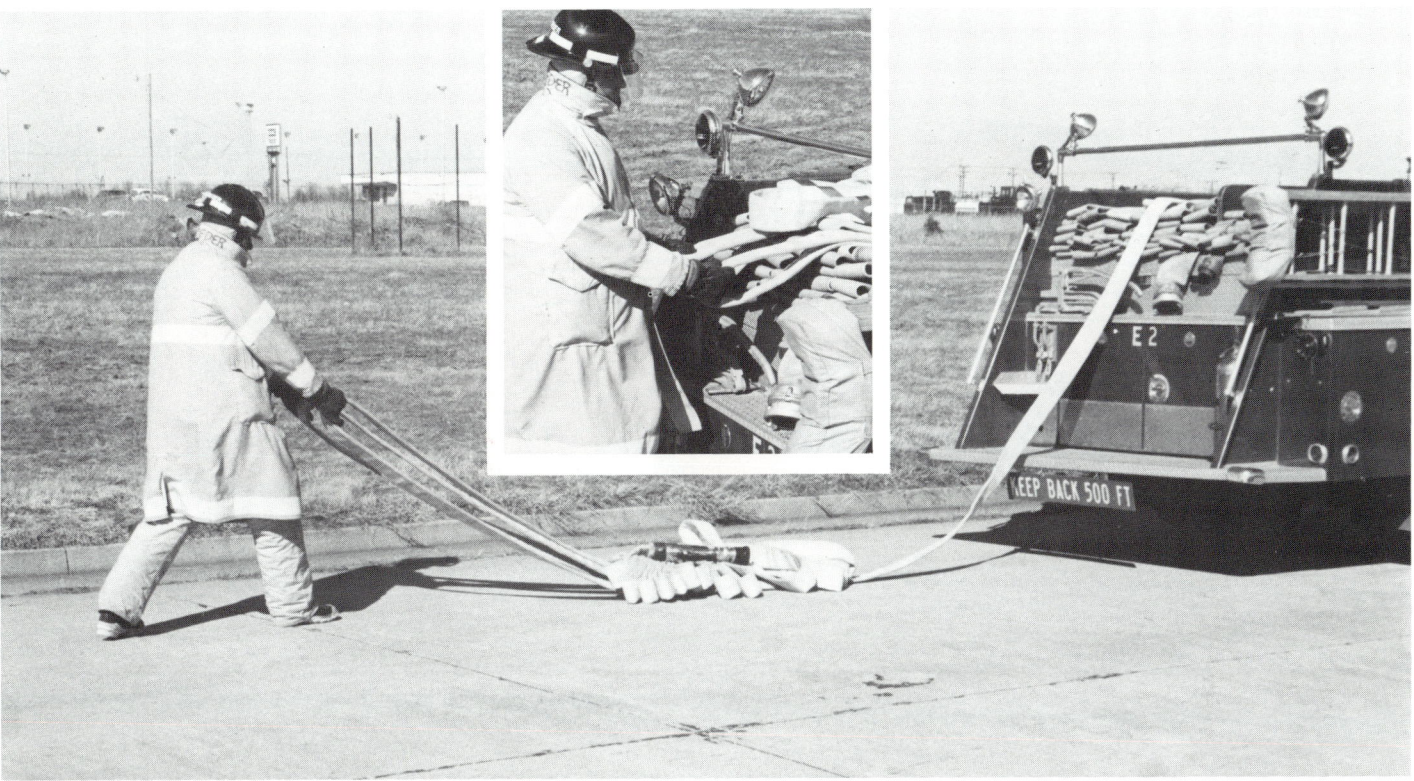

Figure 6.52 STEP 1: Grasp the small loops of each skid and pull the finish from the bed and back from the tailboard approximately 15 feet (5 m).

Figure 6.53 STEP 2: Place a hose clamp on the hose.

it lays out from the bed as the apparatus proceeds toward the water source.

Step 4: After the apparatus makes the lay, pick up the nozzle and lead in the hose.

Step 5: Open the shutoff valve or clamp when ready for water.

PRECONNECTED HOSE LOADS

Another way to carry attack hose is to preconnect it to a discharge valve and place it in an area other than the main hose bed. The primary advantage with preconnecting hose directly to a pump discharge valve is that valuable time is saved when making a quick attack on a fire; an advantage of carrying attack hose outside the main hose bed is that there is no need to move it to the side when laying supply hose, as is often the case with hose finishes.

Any size of small diameter hose can be preconnected to the pump discharge valves. A typical quick-attack operation involves the pumper stopping near the fire, firefighters donning SCBA and pulling the preconnected hoses, and the pump operator charging the hoses with water from the tank as soon as the lines are extended.

190 HOSE

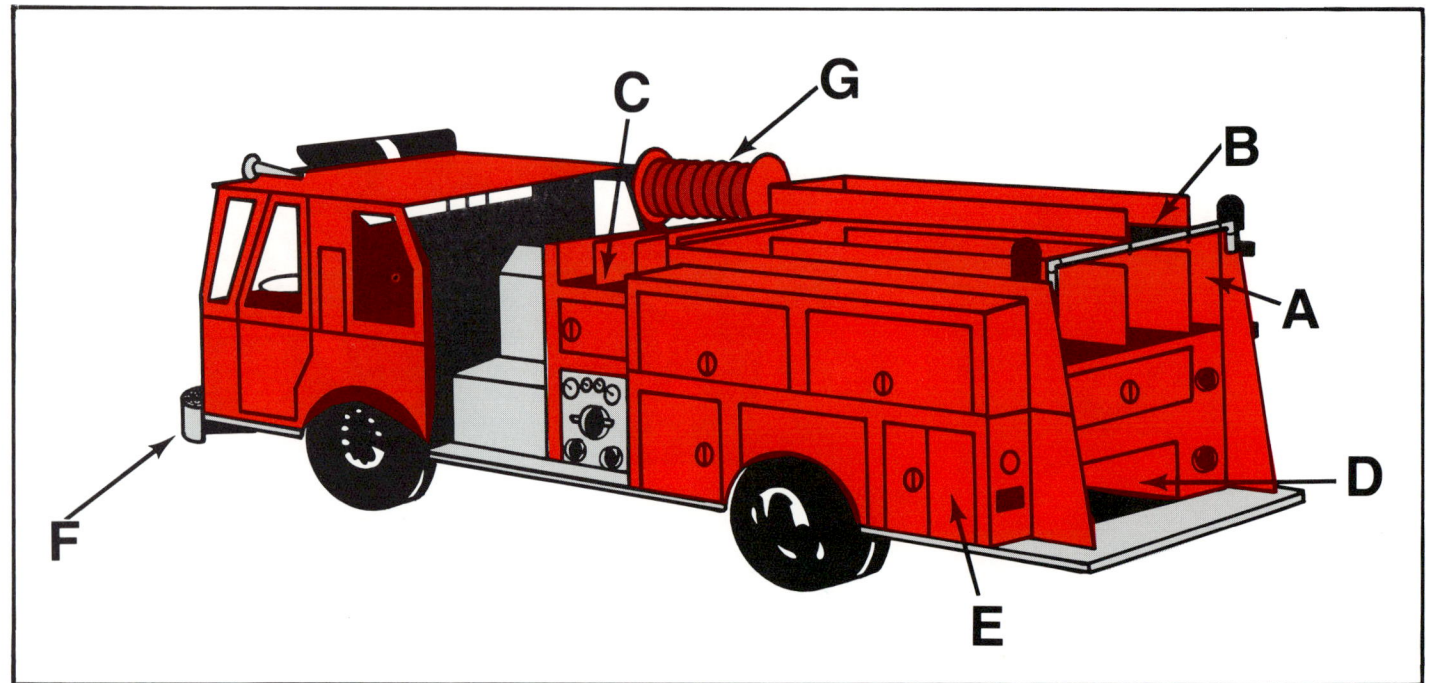

Figure 6.54 Preconnected attack lines can be carried in several places on the apparatus: (A) longitudinal beds; (B) raised trays; (C) transverse beds; (D) tailboard compartments; (E) side compartments or bins; (F) front bumper wells; and (G) reels.

There are several places in which preconnected attack lines can be carried: (Figure 6.54)

- Longitudinal beds
- Raised trays
- Transverse beds
- Tailboard compartments
- Side compartments or bins
- Front bumper wells
- Reels

Longitudinal hose beds for attack hose are generally located alongside the main hose bed, usually to either side. Where space is limited, raised trays mounted over the main hose bed can be used to separate the attack hose from the supply hose beneath. Longitudinal hose beds are a traditional place to carry preconnected hose because, years ago, firefighters were required to ride on the tailboard when responding to fires. Placing hose in beds or raised trays that unload to the rear, therefore, permits firefighters who ride the tailboard to immediately pull the preconnected lines as soon as the apparatus stops at the fire scene. Although many firefighters now ride in jump seats and crew cabs, there is still some advantage in loading attack hose to come off the rear of the engine. Hose that deploys from the rear of the apparatus is in an ideal position when the pumper stops past the fire, a standard practice with some fire departments. Loads that deploy from the rear are also located conveniently close for pulling and disconnecting for a reverse lay. The most notable disadvantage is that the pump operator cannot easily see which lines have been pulled, or if the hose has been completely pulled from the bed.

Transverse hose beds are generally located to the front of the apparatus and deploy to either side. Multiple transverse beds can be placed side by side or can be stacked. The primary advantage with loading preconnected hose into transverse beds is that it places the hose close to seated firefighters. This speeds the quick-attack operation because the firefighters are not required to go to the rear to pull hose. The hose is also less prone to becoming entangled on the apparatus, as sometimes happens when hose is pulled from rear beds and led to the side. It also provides separation between personnel pulling supply lines from the rear of the main hose bed. Another advantage with transverse beds is that the pump operator can more easily see the deployed hose from the pump panel.

Attack hose is sometimes carried in a tailboard compartment to make it more accessible.

Carrying hose in this way has some of the same advantages as carrying it in longitudinal beds: because it deploys from the rear of the apparatus, it is in ideal position when the pumper stops past the fire. It also places the load conveniently close for pulling and disconnecting for a reverse lay but does not take up valuable space in the main hose bed. The primary disadvantage is that the amount of hose that can be loaded is also limited, because space is usually limited in a tailboard compartment. This also decreases the number of possible hose load options, and loading is sometimes a difficult task.

Attack hose can be carried in side compartments or bins mounted above the side compartments. This method has some of the same advantages as transverse beds: it places the hose closer to seated firefighters, which shortens the time needed to get from the seats to the hose location (if the fire is located on the opposite side of the apparatus, however, this causes a greater delay than if the hose were in transverse beds). When pulling hose to the same side, the hose is less likely to snag on the apparatus, and personnel can work in a separate area than those who are pulling rear supply lines. Like hose in tailboard compartments, however, space is limited, so the total length of hose that can be carried is also limited.

Hose carried in front bumper wells is especially well suited for a mobile attack on grass fires and for extinguishing vehicle fires. The hose is easy to reach and is easily reloaded. Space in the well is usually limited, however, so this type of load is usually made with no more than two sections of hose.

Reels are most often used for booster hose but can be used for any size attack hose. Some advantages with hose reels are that there are no bends in the hose and that the hose can be easily laid out and reloaded. A disadvantage with woven-jacket hose loaded on a reel, however, is that the entire length of hose must be laid out before it can be charged. This can cause some delay if the fire is located close to the apparatus.

Reverse Horseshoe Loads

Two methods of loading the reverse horseshoe are described here. The first load is for 2½- or 3-inch (65 mm or 77 mm) hose and is best loaded into a wide hose bed. This load is designed so that a large amount of attack hose can be quickly laid by several persons. The second load is for smaller attack hose (1½- or 1¾-inch [38 mm or 45 mm]) and can be loaded into a narrow longitudinal bed or bin. Unlike the first load, it is designed to be pulled and extended by one person.

MEDIUM HOSE METHOD

Because 2½- or 3-inch (65 mm or 77 mm) hose is used here, this load is best put into a wide hose bed. It is designed so that each tier is pulled and carried by one person. Load no more hose in a tier, therefore, than can be carried by a single firefighter (perhaps two sections of hose). If the load is connected to a discharge at the front of the bed, place an extra length of hose to the side of the first tier to provide enough slack to pull the tier clear of the bed. The procedure shown here starts with the hose connected to a rear discharge:

Step 1: Make a fold on each side of the bed as shown in Figure 6.55. These first folds

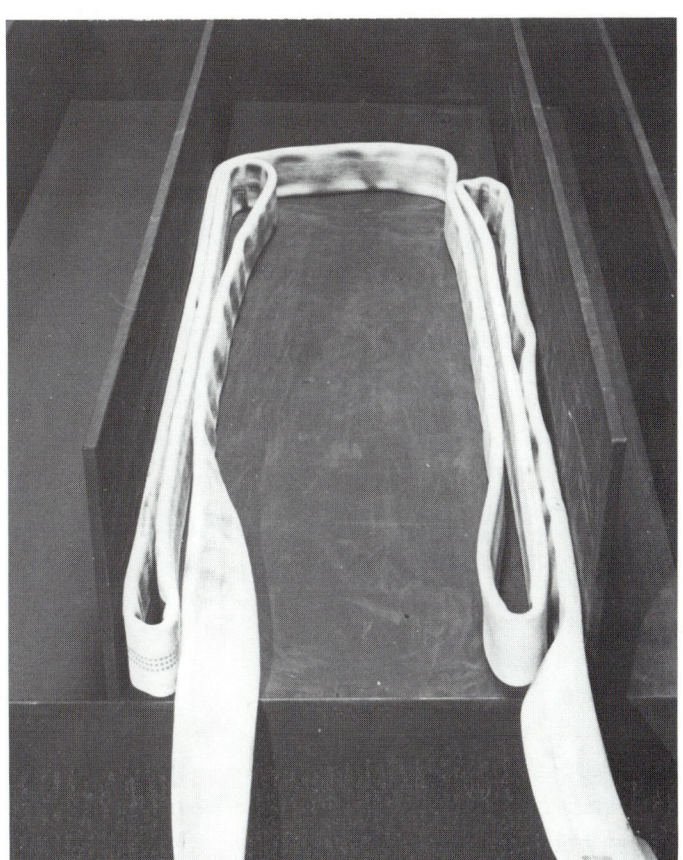

Figure 6.55 STEP 1: Make a fold on each side of the bed.

provide slack when the load is pulled. The hose should go no farther than 5 feet (1.5 m) into the bed (adjust this length so that the hose will not touch the ground when carried over the shoulder).

Step 2: Form the first "U" of the horseshoe by laying the hose around the rear edge of the bed, then return it to the front of the bed and make a fold (Figure 6.56).

Step 3: Lay the next layer of hose inside the previously laid length in the same manner as before (Figure 6.57).

Step 4: Continue to build the horseshoe in the same manner until the first tier is completely filled with hose (Figure 6.58).

Step 5: Load the second and succeeding tiers in the same manner as the first. Start each tier by making a fold of hose on either side of the bed. This will provide slack hose between each person when pulling the hose — one person per tier. Attach the nozzle and place it in the center of the horseshoe loops (Figure 6.59).

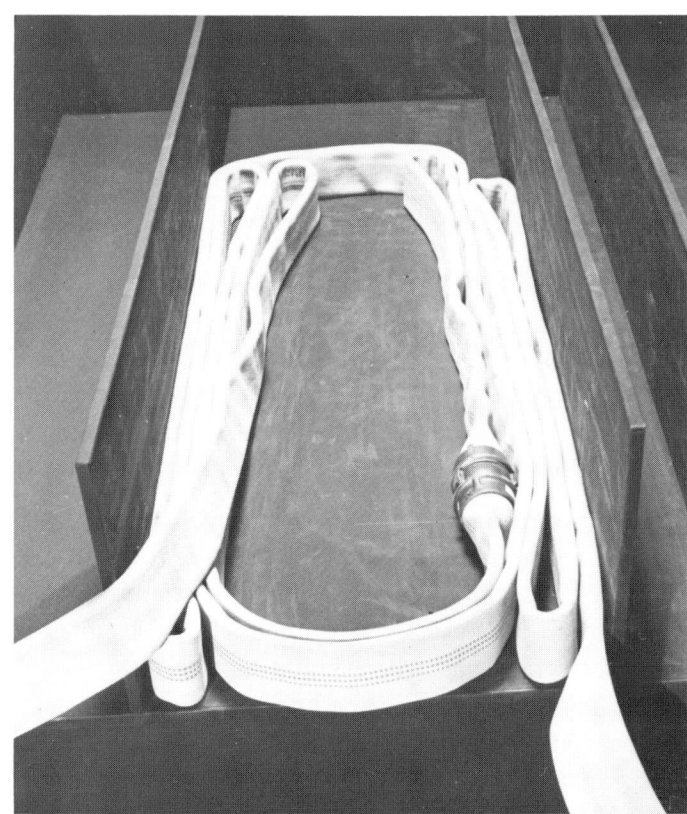

Figure 6.57 STEP 3: Lay the next layer of hose back inside the previously laid length.

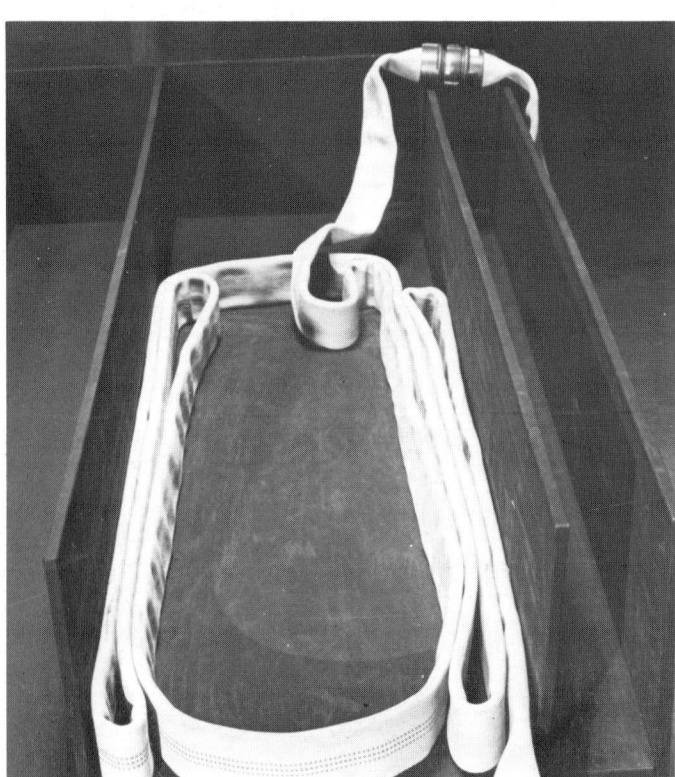

Figure 6.56 STEP 2: Form the first "U" of the horseshoe by laying the hose around the rear edge of the bed, then return it to the front of the bed and make a fold.

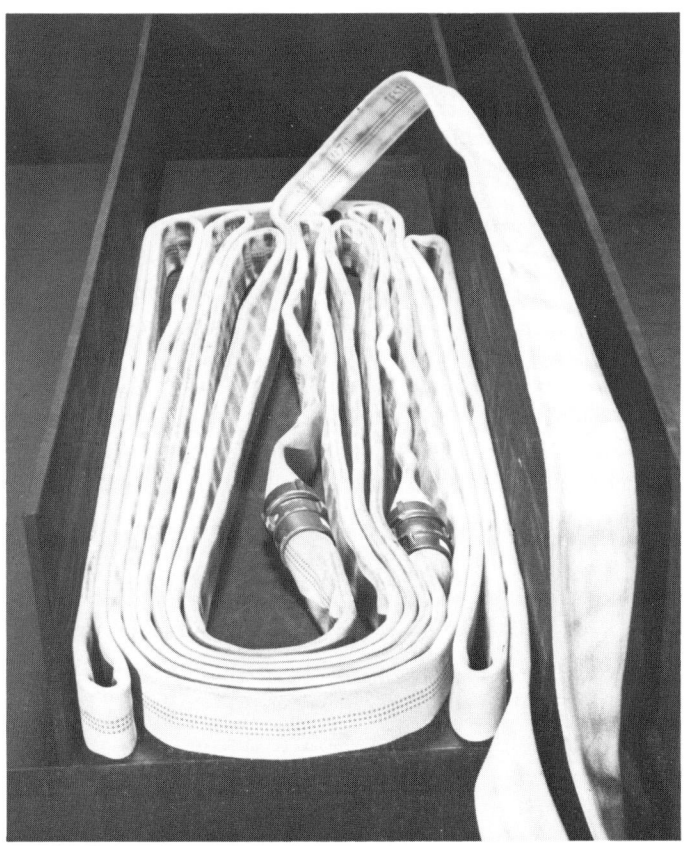

Figure 6.58 STEP 4: Continue to build the horseshoe in the same manner until the first tier is completed.

Attack Hose Loads and Layout Procedures **193**

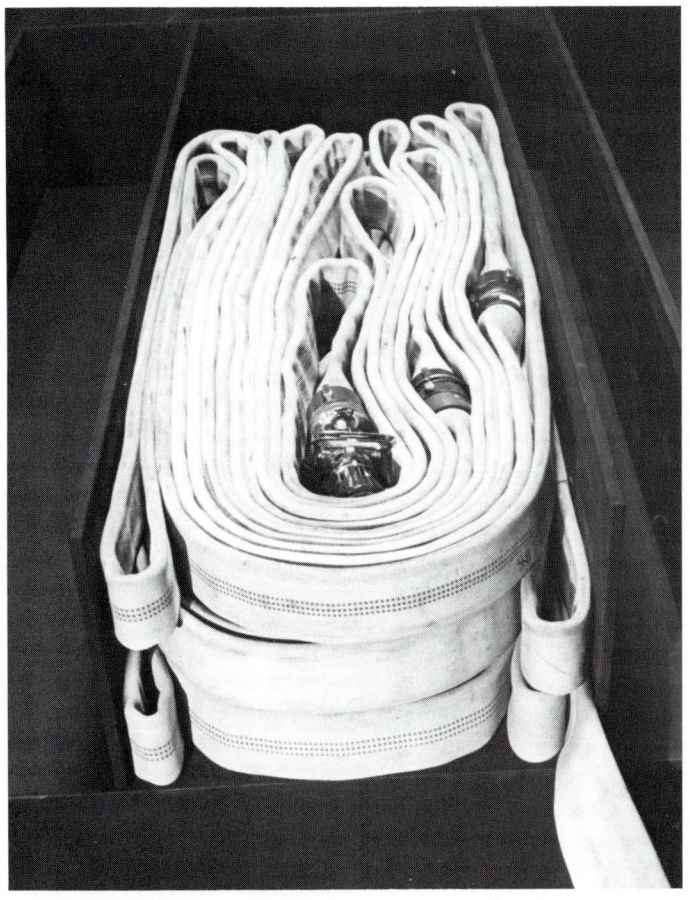

Figure 6.59 STEP 5: Load the second and succeeding tiers in the same manner as the first, then attach the nozzle and lay it in the center of the horseshoe loops.

Figure 6.60 STEP 1: Grasp the nozzle with the outside hand and bring the hose and nozzle behind the head and over the outside shoulder.

Pulling and Advancing

This load is designed so that each person pulls and carries one tier of hose. The hose is carried on the arm so that it pays off as it is carried forward (in the same manner as the accordion shoulder carry, described in Chapter 4). The folds on either side of each tier provide a comfortable separation between firefighters as they lead in the hose. The procedure is as follows:

Step 1: Standing on the tailboard with one shoulder toward the load, grasp the nozzle with the outside hand and bring the hose and nozzle behind the head and over the outside shoulder (Figure 6.60).

Step 2: Pull the loops of the top tier far enough out of the bed to insert the nearest arm, then place that arm through the loops (Figure 6.61).

Step 3: Step down from the tailboard to pull the hose from the bed and walk away from the

Figure 6.61 STEP 2: Pull the loops of the top tier out of the bed and insert the near arm through the loops.

apparatus until the slack hose between the tiers is fully extended (Figure 6.62).

Step 4: Pause so that the second firefighter can secure the second tier in the same manner, then move forward with the second firefighter to extend the slack hose between the second and third tiers (Figure 6.63).

Step 5: When the load is completely pulled from the bed, advance toward the fire (Figure 6.64).

Step 6: Permit the hose to pay off the arm of each firefighter in succession, starting with the rearmost person. Lift each layer from the arm to aid in laying off the hose (Figure 6.65).

Figure 6.64 STEP 5: When the load is completely pulled from the bed, advance toward the fire.

Figure 6.65 STEP 6: Advance toward the fire and lay off the hose from the rearmost firefighter in succession.

Figure 6.62 STEP 3: Pull the first tier from the bed until the slack hose between tiers is fully extended.

Figure 6.63 STEP 4: After the second firefighter pulls the second tier, walk away from the apparatus until the slack hose between the second and third tiers is fully extended.

SMALL HOSE METHOD

This load is similar to the previous method, except that it is for 1½- or 1¾-inch (38 mm or 45 mm) attack hose. The primary advantage with this load is that one person can lay out the hose around obstacles and up stairways without the hose snagging. It is usually loaded into a narrow longitudinal bed or bin so that each tier contains approximately one section of hose. Wider beds may accommodate as much as two sections of hose in each tier. Avoid packing the bed too tightly.

If the load is connected to a discharge at the front of the bed, place an extra length of hose to the side of the first tier to provide enough slack to pull the tier clear of the bed. The procedure shown here

Attack Hose Loads and Layout Procedures **195**

starts with the hose connected to a rear discharge. (A larger amount of hose than normal is used here to fill the wide hose bed):

Step 1: Lay the hose flat down the middle of the bed, then on edge against the side wall (Figure 6.66).

Step 2: Form a "U" just beyond the edge of the bed, then lay the hose to the front against the opposite side. Make a fold and lay the hose back to the rear (Figure 6.67).

Step 3: Lay the hose around the inside of the previously laid length in the same manner as before (Figure 6.68).

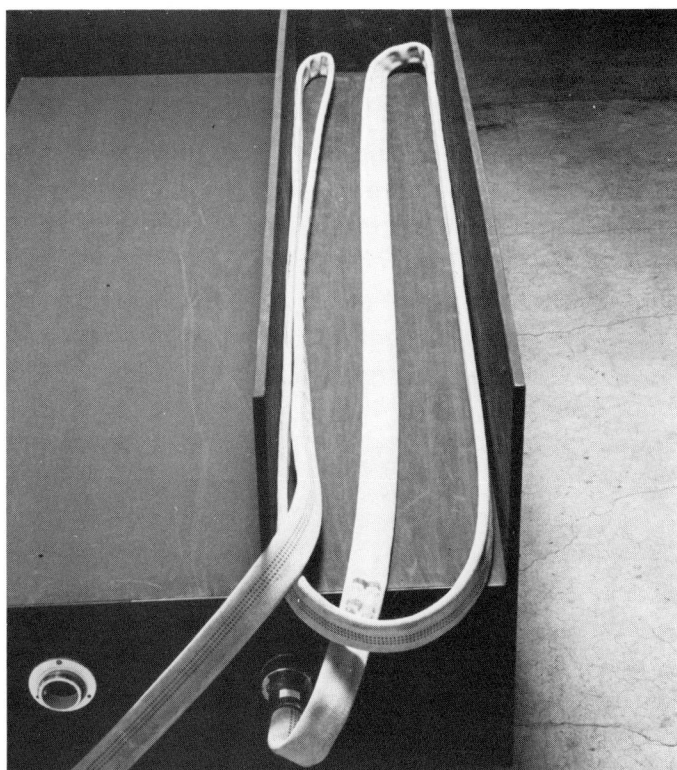

Figure 6.67 STEP 2: Form a "U" just beyond the edge of the bed, then lay the hose to the front against the opposite side. Make a fold and lay the hose back to the rear.

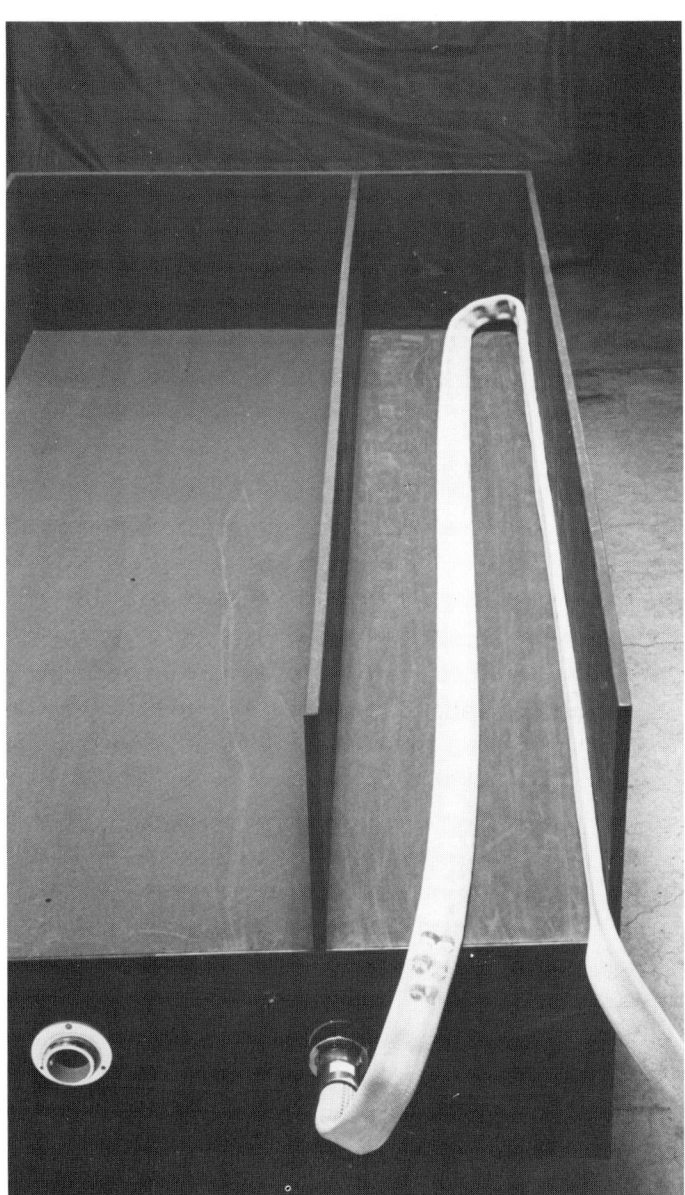

Figure 6.66 STEP 1: Lay the hose flat down the middle of the bed, then on edge against the side wall.

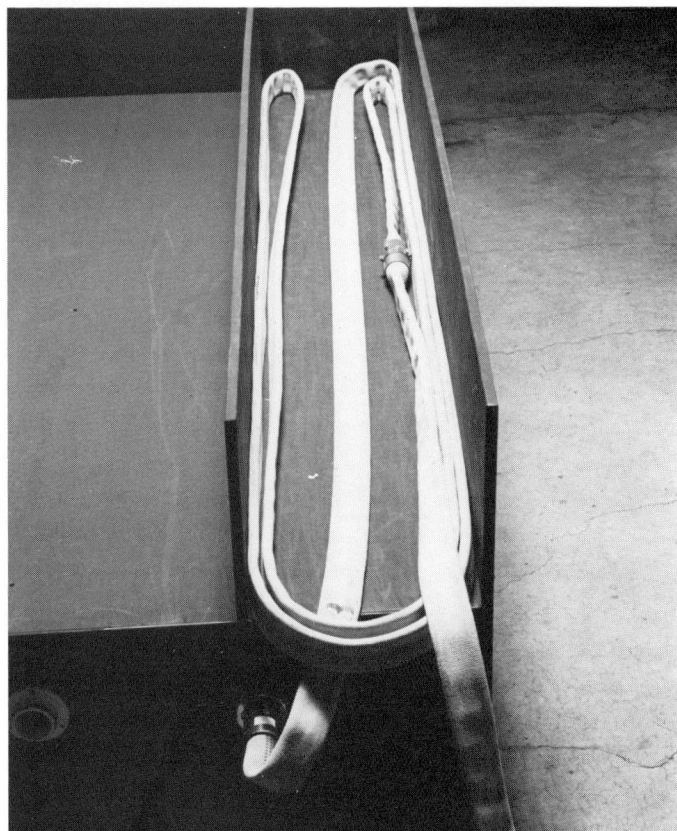

Figure 6.68 STEP 3: Lay the hose around the inside of the previously laid length in the same manner as before.

Step 4: Continue to build the horseshoe in the same manner until the first tier is completely filled with hose (Figure 6.69).

Step 5: Load the second and succeeding tiers in the same manner as the first. Start each tier by bringing the hose up at the front of the bed (from the center of the horseshoe) and across to either side. Attach the nozzle and place it in the center of the horseshoe loops (Figure 6.70).

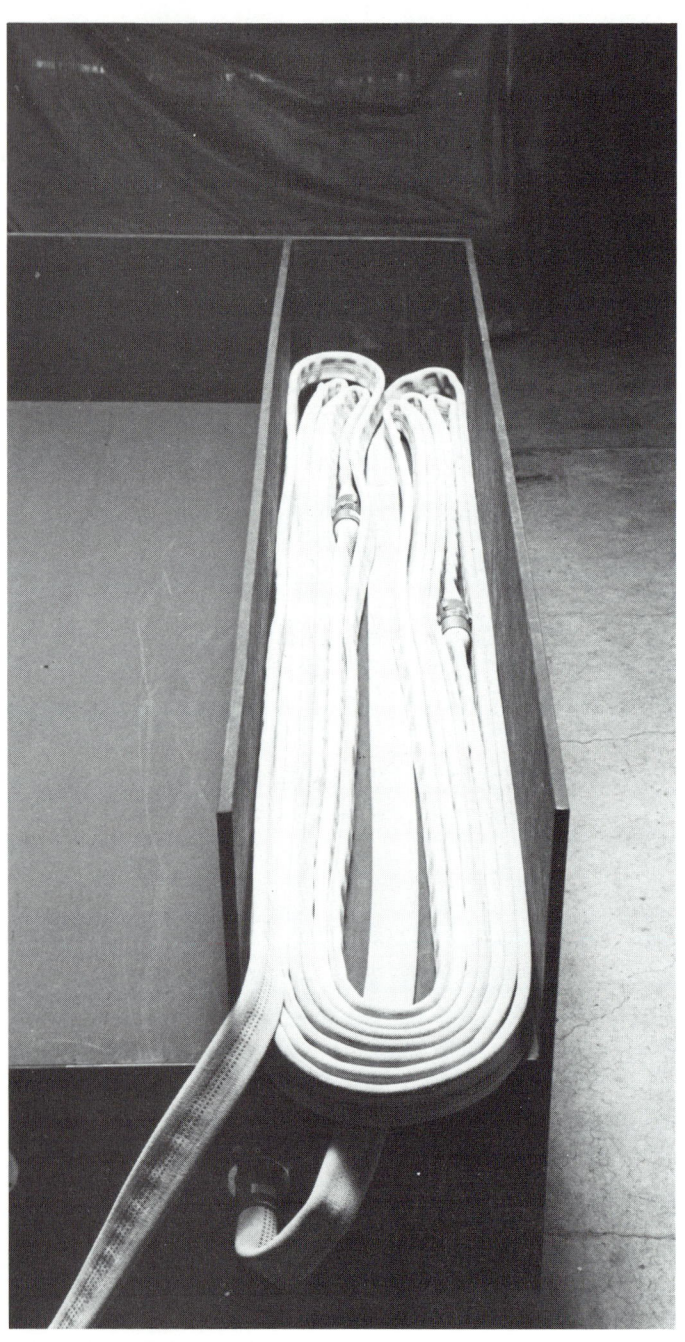

Figure 6.69 STEP 4: Continue to build the horseshoe in the same manner until the first tier is completely filled with hose.

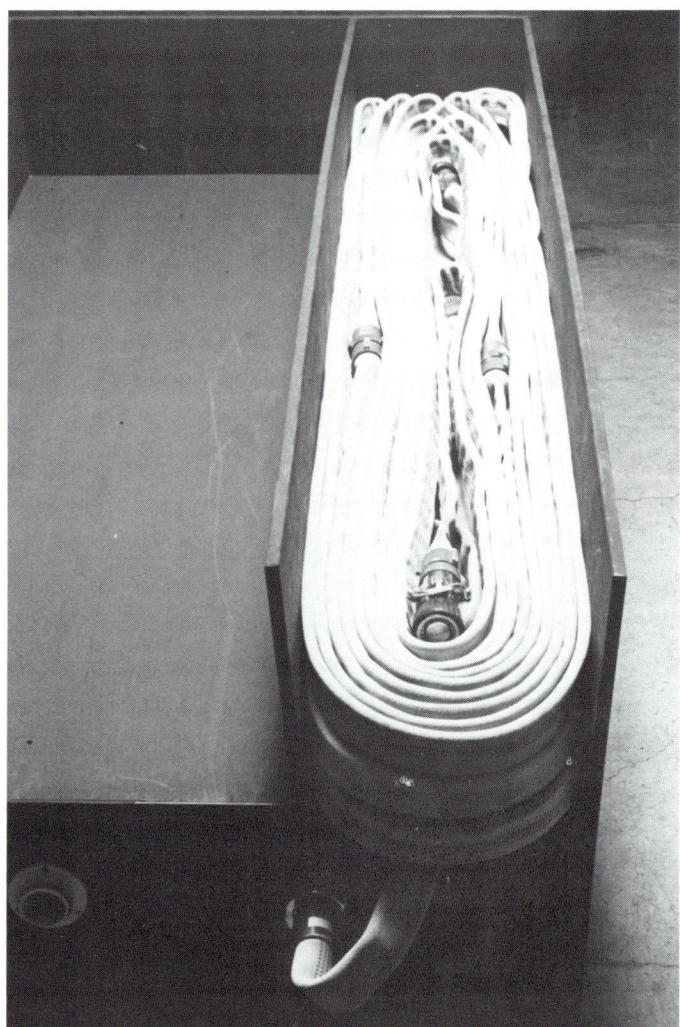

Figure 6.70 STEP 5: Load the second and succeeding tiers in the same manner as the first, then attach the nozzle and place it in the center of the horseshoe.

Pulling and Advancing

Only one firefighter is needed to pull this reverse horseshoe load. The procedure is as follows:

Step 1: Stand on the tailboard facing toward the direction of travel with one shoulder toward the load. Grasp the nozzle with the outside hand and bring the hose and nozzle behind the head and over the outside shoulder so that the nozzle rests in front (Figure 6.71).

Step 2: Put the inside arm over the just-pulled length of hose, then pull the loops of the top tier far enough out of the bed to insert your forearm. Place your forearm through the loops and turn them so that they are perpendicular to the ground (Figure 6.72).

Attack Hose Loads and Layout Procedures **197**

Figure 6.71 STEP 1: Bring the hose and nozzle behind the head and over the outside shoulder so the nozzle rests in front.

Figure 6.72 STEP 2: Pull the hose out far enough to place the arm through the loops. Turn the arm so the loops are perpendicular to the ground.

Figure 6.73 STEP 3: Lift the outside loop of the horseshoe to drop when the hose pulls taut.

Figure 6.74 STEP 4: Lay out the hose one loop at a time.

Figure 6.75 STEP 5: Throw off the remaining loops one at a time to prevent kinking.

each remaining loop off onto the ground in a separate place to eliminate kinking (Figure 6.75).

Step 3: Step down from the tailboard to pull the hose from the bed and begin walking away from the apparatus. With the free hand, lift the outside loop of the horseshoe away from the rest of the loops (Figure 6.73).

Step 4: As the trailing hose tightens, drop the loop and reach for the next loop. Continue to pay off the hose in the same manner until the load is completely deployed (Figure 6.74).

Step 5: If you arrive at your destination with a number of loops still on your arm, throw

198 HOSE

Flat Load

The flat load is adaptable for varying widths of hose beds and is often used in transverse beds. This load is similar to the flat load for larger supply hose (Chapter 5) with two exceptions: it is preconnected, and loops are provided to aid in pulling the load from the bed. The pull loops should be placed at regular intervals within the load so that equal portions of the load are pulled from the bed. The number of loops and the intervals at which they are placed are dependent upon the size and total length of the hose. The procedure shown is for 150 feet (45 m) of 1¾-inch (45 mm) hose loaded into a transverse hose bed:

Step 1: Attach the female coupling to the discharge. Lay the first length of hose flat in the bed against the side wall (Figure 6.76).

Step 2: Angle the hose to lay the next fold adjacent to the first fold. Continue building the first tier in this manner (Figure 6.77).

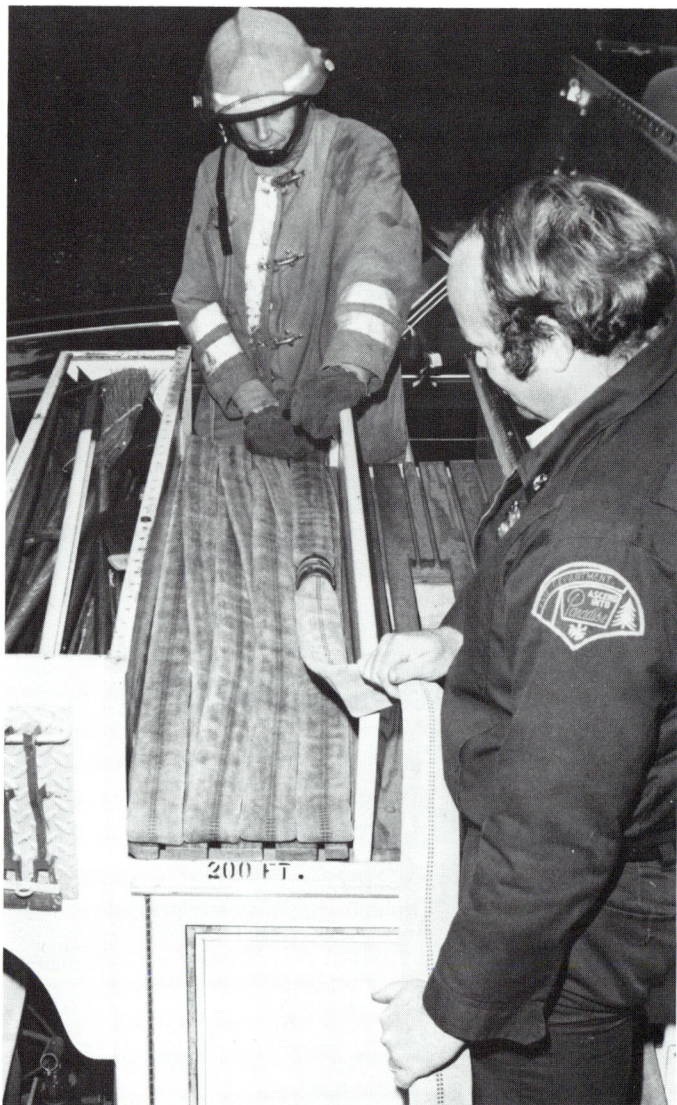

Figure 6.77 STEP 2: Lay the next fold adjacent to the first fold and continue building the first tier in this manner.

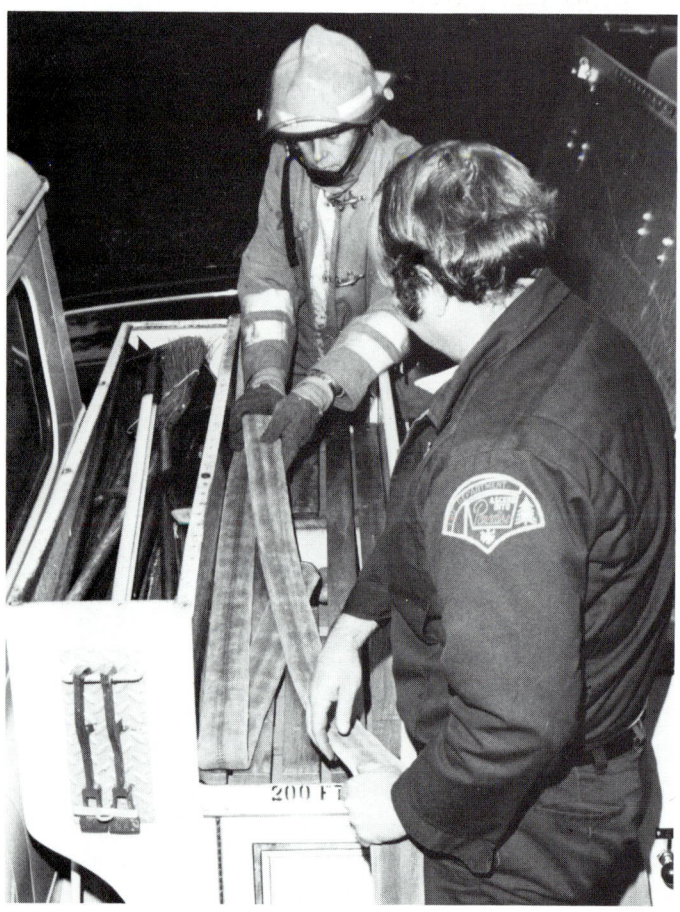

Figure 6.76 STEP 1: Lay the first length of hose flat in the bed against the side wall.

Step 3: At a point approximately one-third of the total length of the load, make a fold that extends approximately 8 inches (203 mm) beyond the load (Figure 6.78). This loop will later serve as a pull handle.

Step 4: Continue laying the hose in the same manner, building each tier with folds laid progressively across the bed. At a point approximately two-thirds of the total length of the load, make a fold that extends approximately 14 inches (356 mm) beyond the load (Figure 6.79). This loop will also serve as a pull handle.

Step 5: Complete the load, then attach the nozzle and lay it on top of the load (Figure 6.80).

Attack Hose Loads and Layout Procedures 199

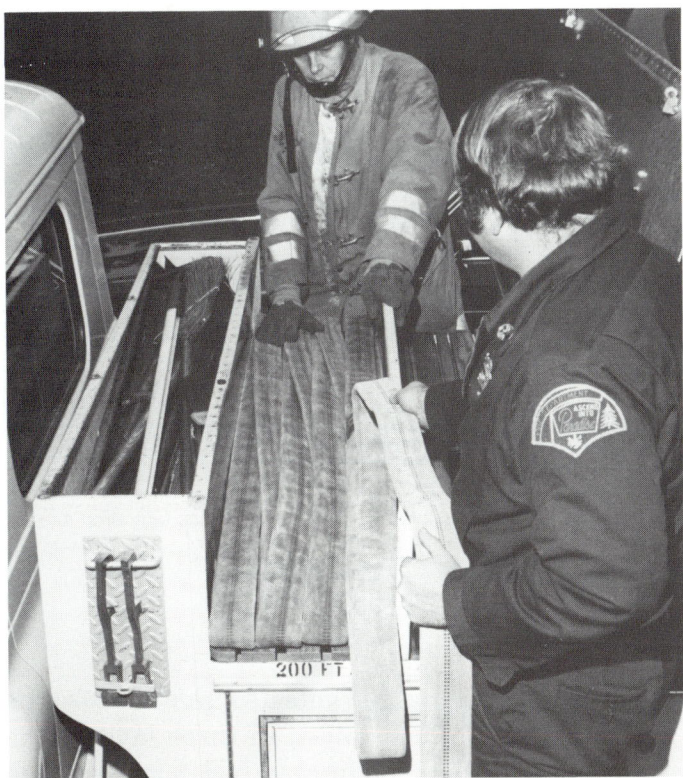

Figure 6.78 STEP 3: Make a pull loop that extends approximately 8 inches (203 mm) beyond the load.

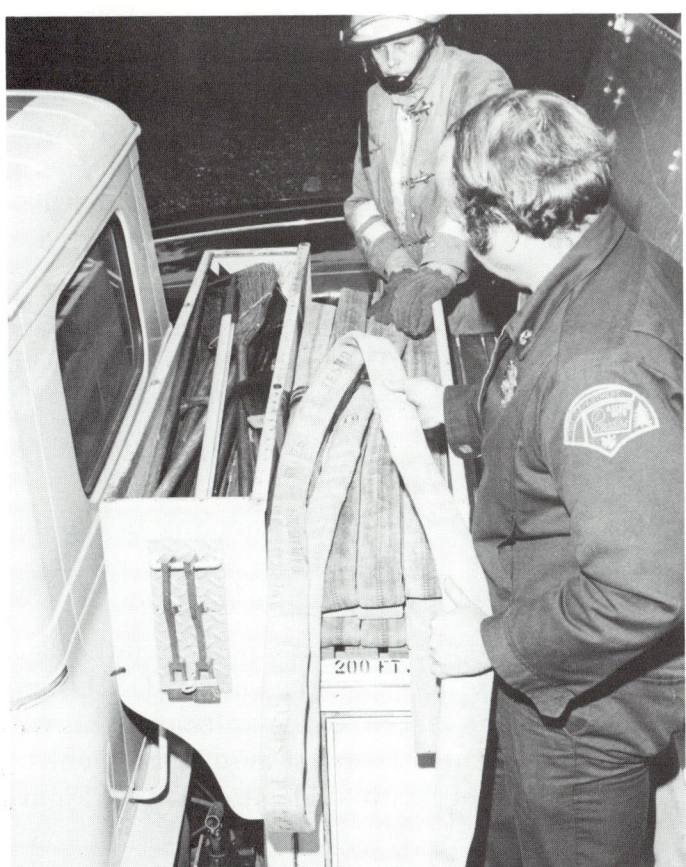

Figure 6.79 STEP 4: Make a second pull loop that extends approximately 14 inches (356 mm) beyond the load.

Figure 6.80 STEP 5: Complete the load, then attach the nozzle and lay it on top of the load.

PULLING AND ADVANCING

The flat load in a transverse bed is generally designed to be pulled and advanced by one person. In beds loaded with an exceptionally large load of hose, two persons may be required to fully extend the load. The procedure shown is for 150 feet (45 m) of 1¾-inch (45 mm) hose. It is pulled by one person:

Step 1: Put one arm through the longer loop and grasp the shorter pull loop with the same hand. Grasp the nozzle with the opposite hand (Figure 6.81).

Figure 6.81 STEP 1: Put one arm through the longer loop and grasp the shorter pull loop with the same hand. Grasp the nozzle with the opposite hand.

Step 2: Using the pull loops, pull the load from the bed (Figure 6.82).

Step 3: Walk toward the fire. As the hose pulls taut in the hand, release the hand loop (Figure 6.83).

Step 4: Continue to lead in the hose. As the shoulder loop becomes taut, drop it to the ground and proceed until the hose is fully extended (Figure 6.84).

Figure 6.82 STEP 2: Using the pull loops, pull the load from the bed.

Figure 6.83 STEP 3: Walk toward the fire. As the hose pulls taut in the hand, release the hand loop.

Figure 6.84 STEP 4: As the shoulder loop becomes taut, drop it to the ground and proceed until the hose is fully extended.

Triple Layer Load

The triple layer load gets its name because the load begins with hose folded in three layers. The three folds are then laid into the bed in an S-shaped fashion. The load is designed to be pulled by one person. A disadvantage with the triple layer load is that the three layers, which may be as long as 50 feet (15 m), must be completely removed from the bed before leading in the nozzle end of the hose. This could be a problem if other apparatus are parked directly behind the hose bed. The load is also relatively cumbersome to place in the hose bed. Four to five persons are usually required to keep the folds aligned as they are laid into the bed.

Start with the sections of hose connected and the nozzle attached. The procedure shown here is for 200 feet (60 m) of 1½-inch (38 mm) hose (the rear discharge has been extended and wyed to supply two preconnected lines).

Step 1: Connect the female coupling to the discharge and extend the hose in a straight line to the rear (Figure 6.85).

Step 2: Pick up the hose at a point two-thirds of the distance from the tailboard to the nozzle and carry it to the tailboard. This will form three layers of hose stacked one on the other with a fold at each end (Figure 6.86).

Step 3: With several persons picking up the entire length of the three layers, begin lay-

Attack Hose Loads and Layout Procedures

Figure 6.85 STEP 1: Connect the female coupling to the discharge and extend the hose in a straight line to the rear.

Figure 6.86 STEP 2: Form three layers of hose stacked one on the other with a fold at each end.

Figure 6.87 STEP 3: Begin laying the hose into the bed by folding over the three layers into the hose bed.

Figure 6.88 STEP 4: Fold the layers over at the front of the bed and lay them back to the rear on top of the previously laid hose. Make all folds at the rear even with the edge of the hose bed.

Figure 6.89 STEP 5: Continue until the entire length is loaded.

ing the hose into the bed by folding over the three layers into the hose bed (Figure 6.87).

Step 4: Fold the layers over at the front of the bed and lay them back to the rear on top of the previously laid hose (if the hose compartment is wider than one hose width, alternate folds on each side of the bed). Make all folds at the rear even with the edge of the hose bed (Figure 6.88).

Step 5: Continue to lay the hose into the bed in an S-shaped configuration until the entire length is loaded (Figure 6.89). If desired, use a rope or strap to secure the nozzle to the first set of loops.

PULLING AND ADVANCING

The triple layer load can be pulled and advanced by one person. The load should be completely pulled from the bed before leading in the nozzle end. The procedure is as follows:

Step 1: Pull the nozzle and folds of the first tier from the bed and, facing away from the load, place them over the shoulder (Figure 6.90).

Step 2: Step down from the tailboard and walk straight away from the rear of the apparatus until the entire load is pulled from the bed (Figure 6.91).

Step 3: Drop the folded end from the shoulder and advance the nozzle to the fire (Figure 6.92).

Figure 6.92 STEP 3: Drop the folded end from the shoulder and advance the nozzle to the fire.

Figure 6.90 STEP 1: Place the nozzle and fold of the first tier over the shoulder.

Figure 6.91 STEP 2: Walk straight away from the rear of the apparatus until the entire load is pulled from the bed.

Minuteman Load

The minuteman load is designed to be pulled and advanced by one person. The primary advantage with this load is that it is carried on the shoulder completely clear of the ground, so it will not snag on obstacles. The load pays off the shoulder as the firefighter advances toward the fire. The load is also particularly well suited for a narrow bed. A disadvantage with the load is that it can be awkward to carry when wearing an SCBA. If the load is in a single stack, it may also collapse on the shoulder if not held tightly in place.

The procedure is for 150 feet (45 m) of 1½-inch (38 mm) hose loaded in a double stack:

Step 1: Connect the first section of hose to the discharge. Do not connect it to the other lengths of hose. Lay the hose flat in the bed to the front, then lay the remaining hose out of the front of the bed to be loaded later (Figure 6.93). (If the discharge is at the front of the bed, lay the hose to the rear of the bed, then back to the front before it is set aside. This provides slack hose for pulling the load clear of the bed.)

Step 2: Couple the two remaining hose sections together and attach a nozzle to the male end. Place the nozzle on top of the first

Attack Hose Loads and Layout Procedures **203**

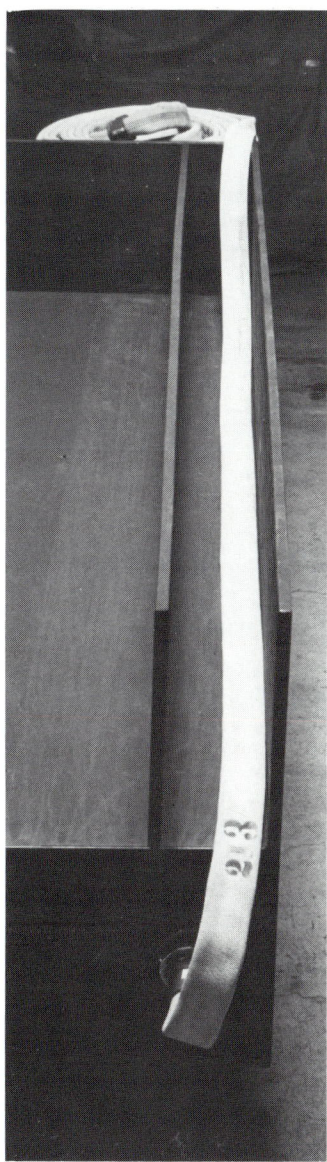

Figure 6.93 STEP 1: Lay the hose flat on one side of the bed to the front, then lay the remaining hose out of the front of the bed to be loaded later.

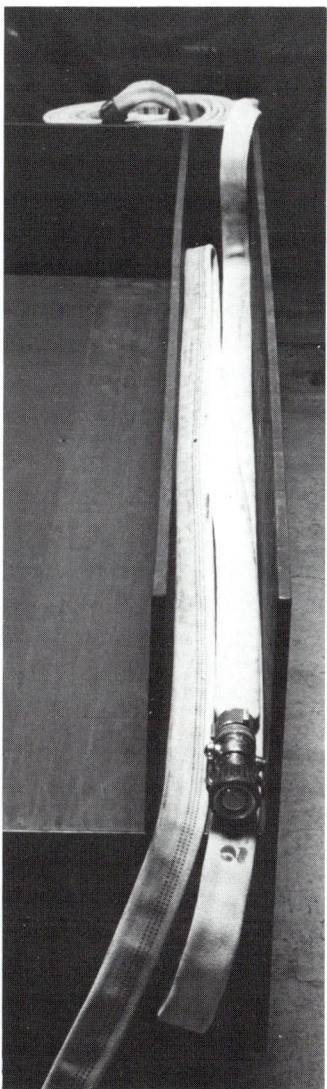

Figure 6.94 STEP 2: Couple the two remaining hose sections together and attach a nozzle to the male end. Place the nozzle on top of the first length at the rear, then angle the hose to the opposite side of the bed and make a fold. Lay the hose back to the rear.

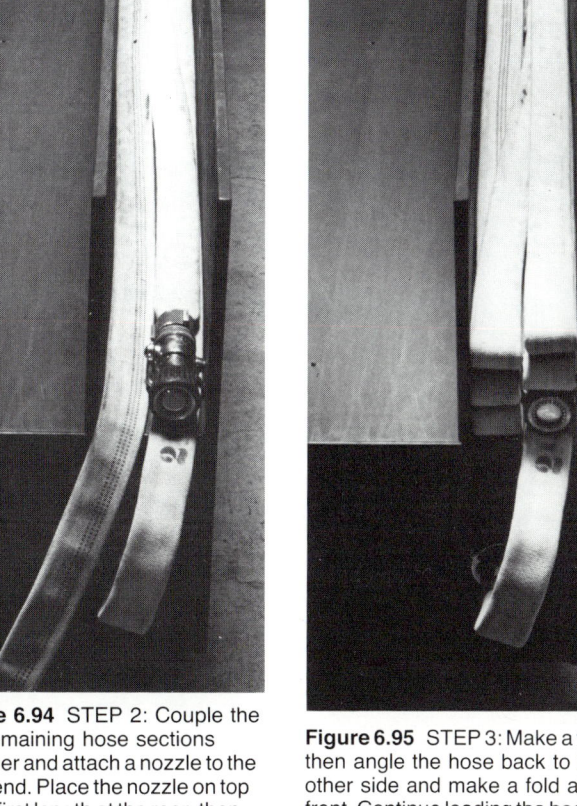

Figure 6.95 STEP 3: Make a fold, then angle the hose back to the other side and make a fold at the front. Continue loading the hose in the same manner to alternating sides of the bed until the complete length is loaded.

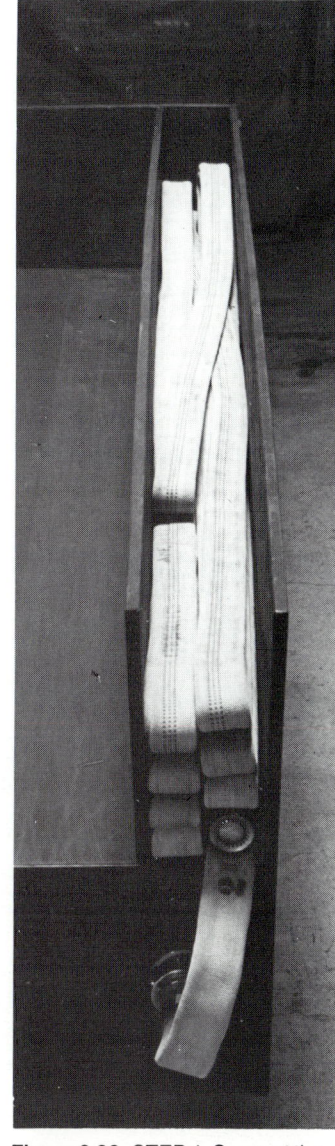

Figure 6.96 STEP 4: Connect the male coupling of the first section to the female coupling of the longer section and lay the remainder of the first section in the bed in the same manner as before.

length at the rear, than angle the hose to the opposite side of the bed and make a fold. Lay the hose back to the rear (Figure 6.94).

Step 3: Make a fold at the rear of the bed, then angle the hose back to the other side and make a fold at the front. Continue loading the hose in the same manner to alternating sides of the bed until the complete length is loaded (Figure 6.95).

Step 4: Connect the male coupling of the first section to the female coupling of the longer section and lay the remainder of the first section in the bed in the same manner as before (Figure 6.96).

PULLING AND ADVANCING

The minuteman load is pulled directly from the bed to the shoulder so that it pays off the shoulder as the load is led in. If the load is in a single stack, it is important to hold the load firmly on the shoulder so that the stack does not collapse, which could cause the load to become entangled. The procedure for leading in the minuteman load is as follows:

Step 1: Grasp the nozzle and pull the load halfway out of the hose bed (Figure 6.97).

Step 2: Face away from the apparatus and place the load on the shoulder with the nozzle length on the bottom. Hold the stack firmly on the shoulder (Figure 6.98).

Step 3: Walk away from the apparatus toward the fire, pulling the remaining hose from the bed (Figure 6.99).

Step 4: Permit the hose to pay off the top of the stack, one fold at a time, as you proceed toward the fire (Figure 6.100).

Figure 6.97 STEP 1: Pull the load halfway out of the hose bed. *Courtesy of Dale Collins.*

Figure 6.98 STEP 2: Place the load on the shoulder with the nozzle length on the bottom. Hold the stack firmly on the shoulder.

Figure 6.99 STEP 3: Walk away from the apparatus toward the fire, pulling the remaining hose from the bed.

Figure 6.100 STEP 4: Permit the hose to pay off the top of the stack, one fold at a time, as you proceed toward the fire.

Wyed Flat Load

The wyed flat load (formerly called the "combination load") consists of two 2½-, 1¾-, or 1½-inch (65 mm, 45 mm, or 38 mm) hoselines connected to a wye. The load shown here is preconnected to a rear discharge, but can be disconnected and attached to 2½- or 3-inch (65 mm or 77 mm) hose laid in a forward or reverse lay (use a double male adapter to connect the wye to hose laid in a forward lay). This procedure is for a wye with 150 feet (45 m) of 2½-inch (65 mm) hose on each side:

Step 1: Connect the wye to the rear discharge, then connect one 150-foot (45 m) section of hose to each side of the wye. Flat lay the first length of one hose to the front of the bed against the side wall (Figure 6.101).

Step 2: Lay the hose back on itself to the rear of the bed, make a fold 10 to 12 inches (254 mm to 305 mm) beyond the hose bed edge, then lay the hose to the front (Figure 6.102). (This loop will be used later to pull the load from the bed.)

Step 3: Continue loading the hose, folding succeeding layers even with the edge of the bed, until the entire length is loaded. Attach a nozzle to complete the load on this side (Figure 6.103).

Step 4: Load the opposite side in the same manner (Figure 6.104).

Figure 6.102 STEP 2: Make a fold 10 to 12 inches (254 mm to 305 mm) beyond the hose bed edge and lay the hose to the front. (Hose is laid temporarily to the side to show the fold.)

Figure 6.103 STEP 3: Continue loading the hose, folding succeeding layers even with the edge of the bed, until the entire length is loaded. Attach a nozzle to complete the load on this side.

Figure 6.101 STEP 1: Flat lay the first length of one hose to the front of the bed.

Figure 6.104 STEP 4: Load the opposite side in the same manner.

PULLING AND ADVANCING

Each side of the wyed load is designed to be pulled and advanced by one person. The procedure shown is for using the load directly as a preconnected load:

Step 1: Standing on the tailboard facing the hose load, grasp the nozzle with one hand and place the opposite hand through the bottom loop (Figure 6.105).

Step 2: Step off the tailboard to pull the layers from the bed (Figure 6.106).

Step 3: Drop the load to the ground and lead in the hose (Figure 6.107).

Step 4: Pull and advance the second side in the same manner as the first (Figure 6.108).

Figure 6.106 STEP 2: Step off the tailboard to pull the layers from the bed.

Figure 6.107 STEP 3: Drop the load to the ground and lead in the hose.

Figure 6.105 STEP 1: Grasp the nozzle with one hand and place the opposite hand through the bottom loop.

Figure 6.108 STEP 4: Pull and advance the second side in the same manner as the first.

HOSE PACKS

One of the best ways to store hose in a small space, such as a side compartment, is in a pack. Another advantage with a hose pack is that it is easily transported from the apparatus to a distant fire location by one or two persons (this depends on the size, length, and type of hose, as well as the way the pack is designed). Two primary uses for packs are for connecting to standpipes in high-rise buildings and for laying out hose in wildland fire fighting situations.

Standpipe Pack

One of the greatest problems in fighting fires in large buildings, particularly in multistoried buildings, is transporting water to the seat of the fire. Years ago, firefighters were required to lay hose into the buildings and up stairwells to reach the fire floor. Eventually, however, new buildings were equipped with standpipes so that firefighters only needed to transport enough hose to connect to the standpipe and reach the fire. These standpipe systems may be vertical (for example, in high-rise buildings and parking garages) or horizontal (in shopping malls, factories, and warehouses).

Although transporting hose to a standpipe connection is a much simpler evolution than laying hose the entire distance from the pumper to the fire, it can still be an exhausting job. Firefighters must take not only hose and nozzles to the fire location but also SCBA, spare bottles, and tools. With this in mind, lightweight hose is often packed into portable packs to make it more compact and thus more transportable.

There are many types of packs: some are made to be carried on the firefighter's back, some are hand carried, and others are secured to two-wheeled hand carts. Figure 6.109 shows the components of a typical hand-carried standpipe pack. This kit contains the following equipment:

- A lightweight tubular frame with nylon straps for securing the hose to the frame. The frame has a built-in handle so that one or two persons can carry the pack.

- A 100-foot (30 m) length of 1½-inch (38 mm) 800 psi (5 520 kPa) proof test lightweight hose.

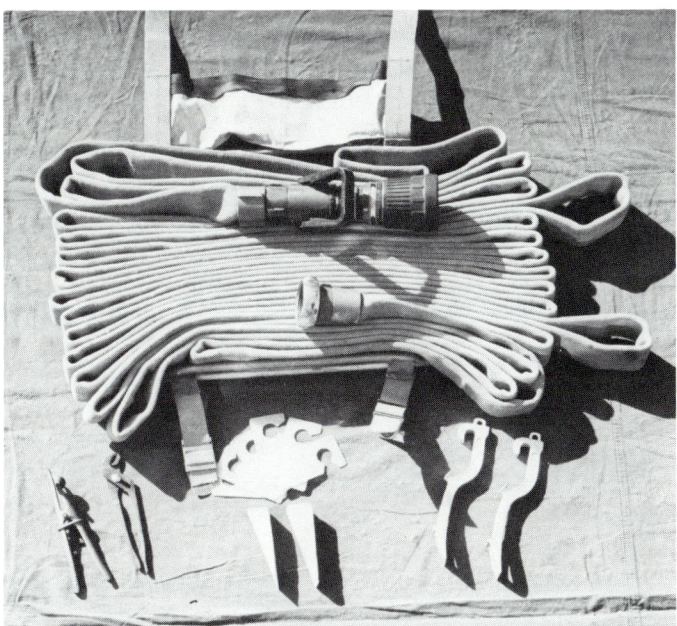

Figure 6.109 Components of a typical portable standpipe pack.

- A 2½ x 1½-inch (65 mm by 38 mm) reducer.
- An automatic fog nozzle.
- A canvas pouch (with hook-and-loop closures) containing sprinkler tongs and wedges, channel lock pliers, lightweight spanner wrenches, and search and rescue tags.

The pack is accordion loaded in a single stack, as shown in Figure 6.110. If carried by two persons,

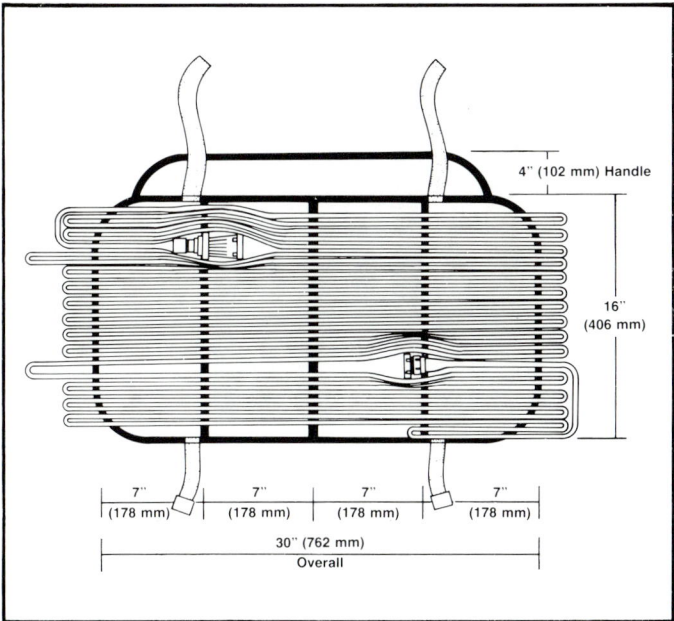

Figure 6.110 Hose in this high-rise pack is loaded accordionlike in a single stack.

the total amount of hose can be doubled (loaded in two stacks). Figure 6.111 shows a crew carrying the high-rise pack, along with other essential equipment.

When connecting the hose in the standpipe pack to the standpipe, stop one floor below the fire floor to make the connection to the standpipe. Remove the standpipe valve cap and check the opening to make sure there is no debris in the passageway (Figure 6.112), then connect the hose to the standpipe fitting (Figure 6.113). Loops built into the hose pack divide it into thirds so that as one firefighter connects the load to the standpipe, two other firefighters can advance the hose. In this case, the leading firefighter takes the nozzle and first loop and leads in; the second firefighter takes the second loop and advances the balance of the load (Figure 6.114).

Lay the hose to the outside of the stairway to prevent it from kinking at the stairwell turns when charged (Figure 6.115). If fire conditions permit, lay the hose up the stairs toward the floor *above* the fire. This will make it easier to advance a charged hoseline, as the hose can be pulled down the stairs rather than having to be pulled up from the stairway below.

Figure 6.111 This crew is carrying a high-rise pack and other essential equipment into a high-rise building.

This series of photos (6.109-6.115) Courtesy of West Palm Beach Florida Fire Department.

Figure 6.112 Before connecting to the standpipe outlet, check for debris in the opening.

Figure 6.113 Connect the hose to the standpipe fitting.

Figure 6.114 The leading firefighter takes the nozzle and first loop and leads in; the second firefighter takes the second loop and advances the balance of the load.

Attack Hose Loads and Layout Procedures 209

Figure 6.115 Lay the hose to the outside of the stairway to prevent the hose from kinking at the stairwell.

Open the standpipe valve and allow the air to discharge through the nozzle. Flow a small amount of water to make certain that the riser is free of debris and that the nozzle is working properly.

CAUTION: Be aware that directing a fog stream into a highly heated, unvented room from a doorway sometimes creates a situation where steam is quickly generated and pressurizes the area. This pressure can push the fire out the door to surround personnel in the stairwell.

Wildland Pack

The wildland pack is an excellent way to transport hose across difficult terrain because it is lightweight, compact, and easy to deploy. Its backpack design, which uses part of the hose as shoulder straps (Figure 6.116), frees the carrier's hands for climbing and for carrying tools and other equipment.

This pack is made of two 100-foot (30 m) sections of 1½-inch (38 mm) single-jacket hose. The sections are tied together with lightweight cord in such a manner that one section at a time can be dropped from the pack and deployed. Eight 52-inch (132 mm) cords are needed to tie the pack together. (These cords should be made from parachute shroud line for maximum strength and durability.) The cords are each prepared with a fixed loop at one end so that slipknots can be made when the sec-

Figure 6.116 The wildland pack uses part of the hose as shoulder straps to free the carrier's hands for climbing and for carrying tools and other equipment. *Courtesy of Darren Read, Oroville, California.*

210 HOSE

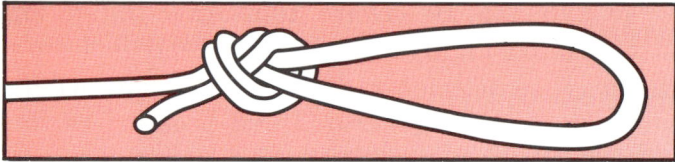

Figure 6.117 Make a fixed loop at one end of the tie cord to permit the making of a slipknot when tying the hose pack.

tions of hose are tied into place (Figure 6.117). It is critical to tie the slipknot so that the cord pulls clear of the hose when the hose is deployed. Figure 6.118 illustrates the proper method for tying the slipknots. Use pliers to pull the cords tightly around the bundles.

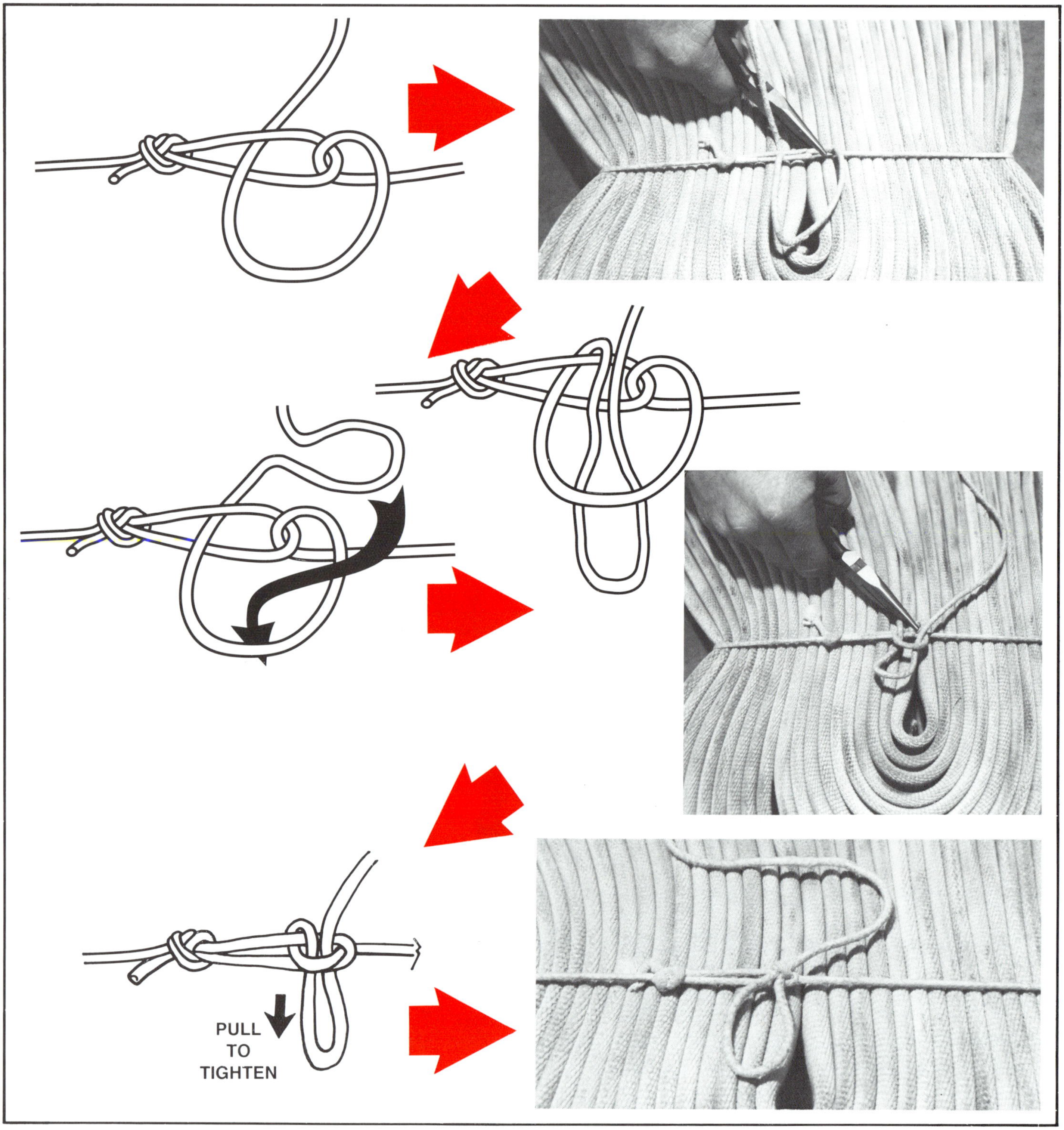

Figure 6.118 Make the slipknot so that it comes completely free of the hose pack when pulled.

Attack Hose Loads and Layout Procedures **211**

The wildland pack can be made on any flat, clean surface, but is most easily started on a table specifically designed for making the coils in the appropriate diameter. Figure 6.119 illustrates the table and its dimensions. The procedure for making the wildland pack is as follows:

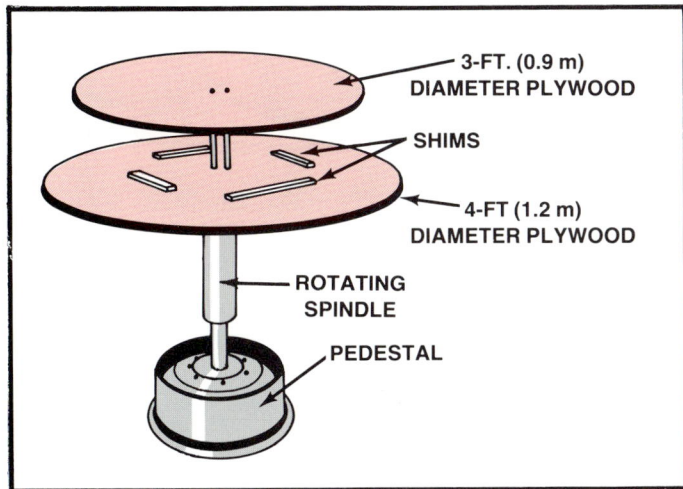

Figure 6.119 Custom-built table for building wildland packs. *Courtesy of Chico, California Fire Department.*

Step 1: Starting with the male end, lay the first section of hose tightly around the center form (a 3-foot [0.9 m] diameter circle) (Figure 6.120).

Step 2: Maintaining the tight coils, lift the hose from the table and place it on the floor (Figure 6.121).

Step 3: Push the coils into a horseshoe shape so that the female coupling is at the bottom

Figure 6.121 STEP 2: Lift the hose from the table, maintaining the tight coils, and place it on the floor.

and center of the bundle, then place tie cords under the bundle. Bring the ends of the top cord inside the first layer of hose (this layer will become the shoulder loops for carrying the wildland pack) (Figure 6.122).

Step 4: Tie the top cord tightly with a slipknot that lies over the center of the bundle, leaving the outside layer of hose free to serve as shoulder straps (Figure 6.123 on next page).

Figure 6.120 STEP 1: Lay the first section of hose tightly around the center form.

Figure 6.122 STEP 3: Push the coils into a horseshoe shape with the female coupling at the bottom center of the bundle. Place tie cords under the bundle and bring the ends of the top cord inside the first layer of hose.

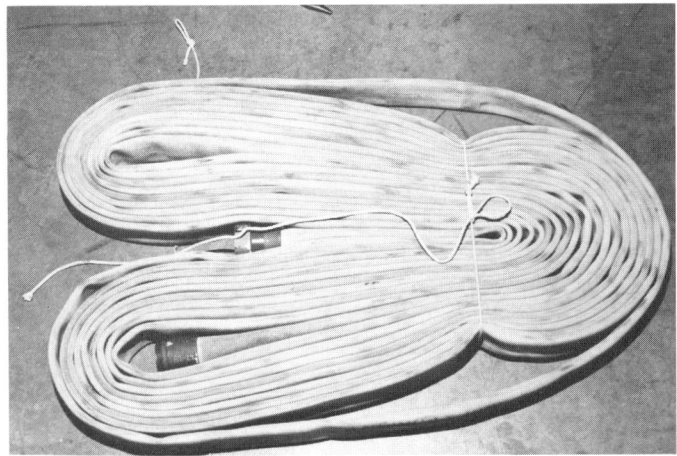

Figure 6.123 STEP 4: Tie the bundle tightly at the top of the horseshoe with a slipknot, leaving the outside layer of hose free to serve as shoulder straps.

Step 5: Tie the second cord across the bottom part of the horseshoe with a slipknot, then tie the trailing ends of this cord and the first cord with a square knot (Figure 6.124).

Step 6: Tie the top center cord of the horseshoe in the same manner as before (over all the layers), then tie its trailing end to one of the first two cords with a square knot (Figure 6.125).

Step 7: Make the second bundle in the same manner as the first, but do not make shoulder loops (include the outside layer of hose in the second tie) (Figure 6.126).

Step 8: Lay the second bundle on top of the first bundle so that the pull cords of each bundle are to the inside of the pack (Figure 6.127).

Step 9: Bind the two bundles together with two cords tied with slipknots, one at the top

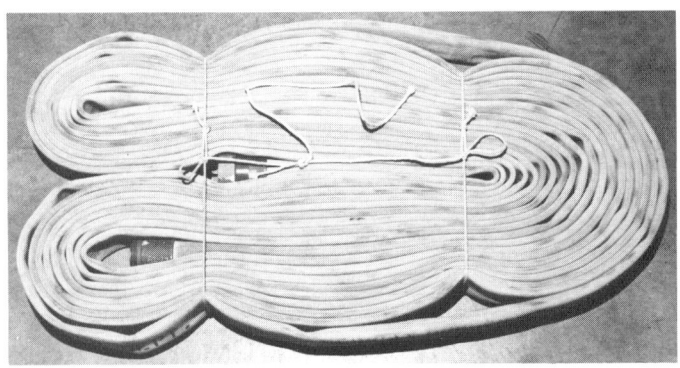

Figure 6.124 STEP 5: Tie the second cord with a slipknot, then tie the two cords together with a square knot.

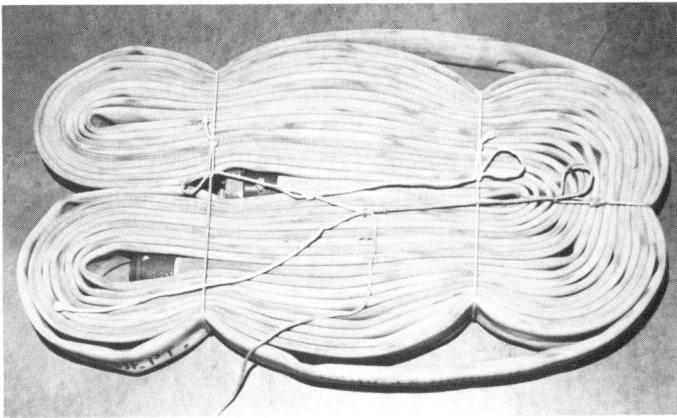

Figure 6.125 STEP 6: Tie the third cord with a slipknot over all the layers of the hose, then tie its trailing end to one of the first two cords with a square knot.

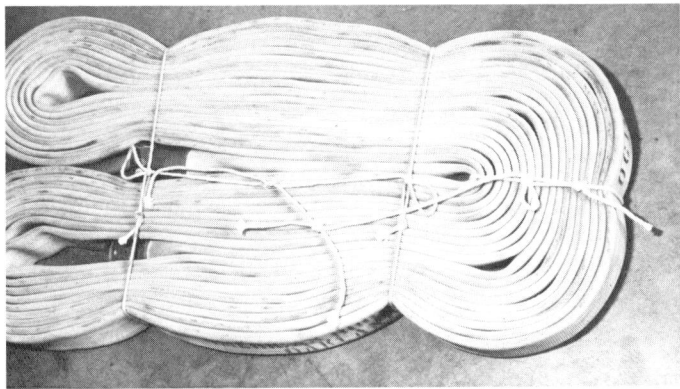

Figure 6.126 STEP 7: Make the second bundle in the same manner as the first, but do not make shoulder loops.

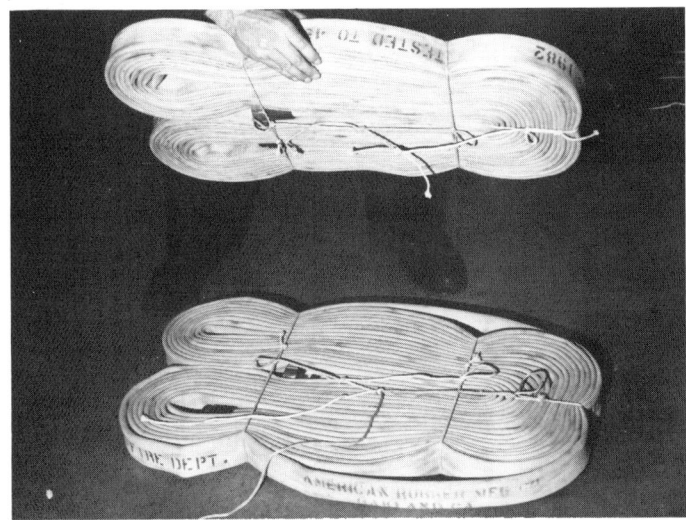

Figure 6.127 STEP 8: Lay the second bundle on top of the first bundle so that the pull cords of each bundle are to the inside of the pack.

and the other at the bottom of the bundle. Then tie the ends of the two cords together with a square knot to make a pull cord (Figure 6.128).

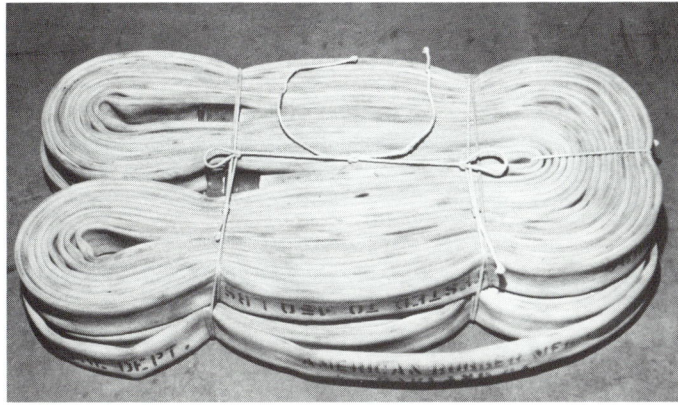

Figure 6.128 STEP 9: Bind the two bundles with two cords slipknotted, then tie the ends of the two cords together with a square knot to make a pull cord.

PROGRESSIVE HOSE LAY FROM A WILDLAND PACK

The wildland pack is specifically designed to not only transport hose across difficult terrain, but also to progressively lay out this hose during a direct attack on the fire. Firefighters involved in progressive hose lays usually carry an assortment of tools that include hose clamps, nozzles, in-line tees, and adapters (Figure 6.129). This procedure is for carrying and progressively laying out hose in the wildland pack:

Step 1: Pull the shoulder loops free of the bundle and move the top layer of hose beneath the center tie to the front of the pack (Figure 6.130).

Step 2: Stand the pack in an upright position with the shoulder loops at the bottom. Grasp the sides of the pack with your hands inside the loops (Figure 6.131).

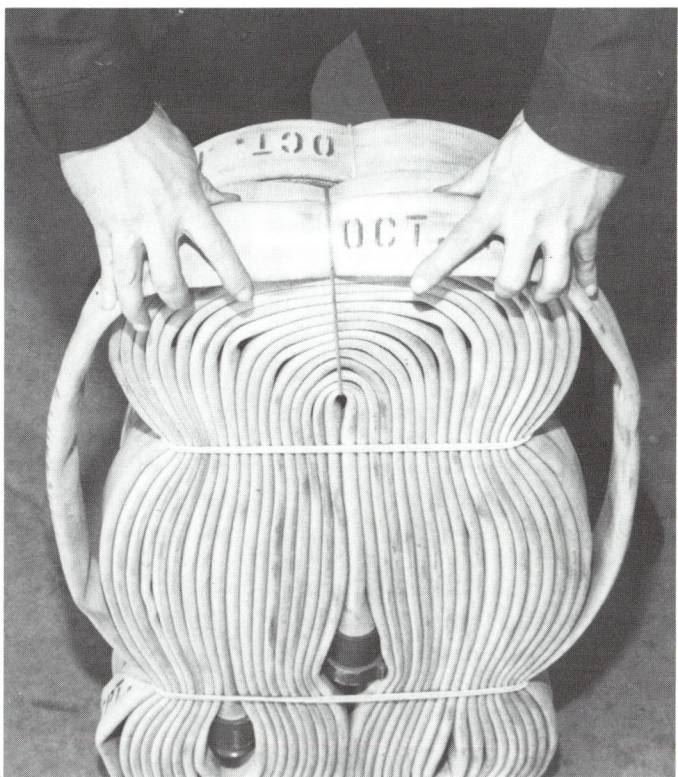

Figure 6.130 STEP 1: Pull the shoulder loops free of the bundle.

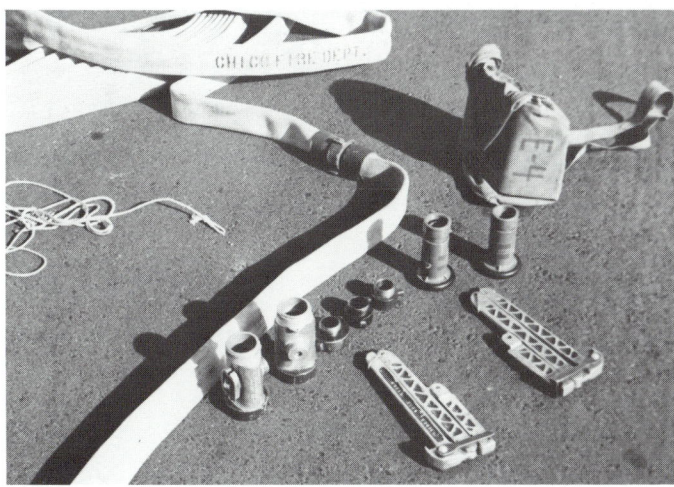

Figure 6.129 A wildland tool pack contains hose clamps, nozzles, in-line tees, and adapters.

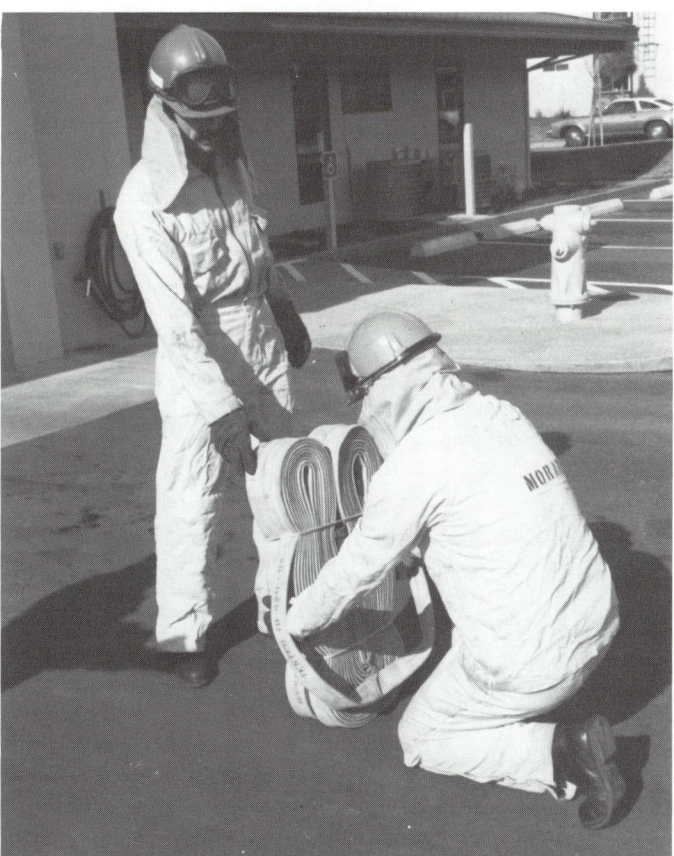

Figure 6.131 STEP 2: Stand the pack in an upright position with the loops at the bottom and grasp the sides of the pack inside the shoulder loops. *Courtesy of Don Moran.*

214 HOSE

Step 3: Swing the bundle up over your head and gently lower it onto your back with your arms through the shoulder loops (Figure 6.132).

Step 4: Walk to the desired location. At the place where the hose lay starts (point of connection), have a second person sharply yank the outside pull cord. This pulls both slipknots free at the same time to drop the outside bundle to the ground (Figure 6.133).

Step 5: Sharply yank the pull cord on the dropped bundle to release the slipknots on the three cords (Figure 6.134).

Step 6: Push the bundle coils out into their original circular form (Figure 6.135). This will prevent the hose from kinking when charged.

Figure 6.133 STEP 4: Sharply yank the outside pull cord to drop the outside bundle to the ground.

Figure 6.132 STEP 3: Swing the bundle up over your head and gently lower it onto your back with your arms through the shoulder loops.

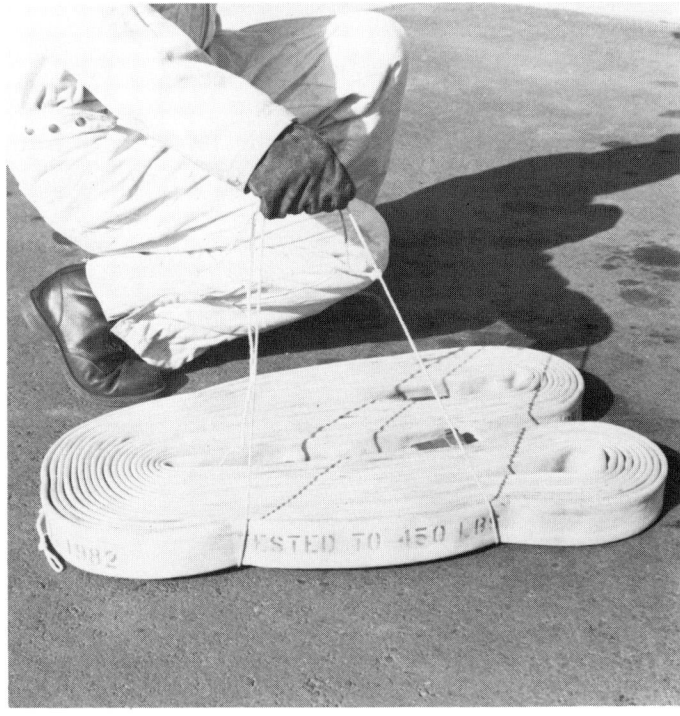

Figure 6.134 STEP 5: Sharply yank the pull cords on the dropped bundle to release the slipknots on three cords.

Attack Hose Loads and Layout Procedures **215**

Figure 6.135 STEP 6: Push the bundle coils out into their original circular form.

Figure 6.137 STEP 8: Remove the nozzle and connect the bundle hose to the working line. Attach the nozzle to the male end of the bundle hose.

Step 7: Clamp the working hoseline near the nozzle, then open the nozzle to release the pressure (Figure 6.136).

Step 8: Remove the nozzle from the working line and connect the bundle hose to the working line. Attach the nozzle to the male end of the bundle hose (Figure 6.137).

Step 9: Release the hose clamp and open the nozzle to expel the air (Figure 6.138).

Step 10: Proceed forward and apply water to the fire. When the entire length of hose from the first bundle is extended, close the nozzle. Drop the second bundle to the ground (Figure 6.139) and repeat Steps 3-9 to lay out the hose.

Figure 6.138 STEP 9: Release the hose clamp and open the nozzle to expel the air.

Figure 6.136 STEP 7: Clamp the working hoseline near the nozzle, then open the nozzle to release the pressure. *Courtesy of John Probst.*

Figure 6.139 STEP 10: Drop the second bundle and repeat the same procedure as before to lay out this hose.

Chapter 6 Review
Answers on page 237

MULTIPLE CHOICE: Circle the correct answer.

1. Which type of preconnected hose load must be completely removed from the bed before leading in the nozzle end of the hose?
   A. Horseshoe
   B. Minuteman
   C. Triple Layer
   D. Wyed Flat

2. On what floor of a building should hose from a standpipe pack be connected to the standpipe?
   A. Ground floor
   B. Floor below the fire
   C. Fire floor
   D. Floor above the fire

TRUE-FALSE: Mark each statement true or false. If false, explain why.

3. Attack hose should be loaded so that it can be pulled and deployed in the most expedient manner.
   ☐ T   ☐ F _____
   _____

4. When a finish is used in a forward lay, the first step in laying the hose is to pull the finish from the hose bed.
   ☐ T   ☐ F _____
   _____

5. When a finish is used in a reverse lay, the first step in laying the hose is to pull the finish from the hose bed.
   ☐ T   ☐ F _____
   _____

6. When using a donut roll as a finish, the female coupling should be on the outside of the roll.
   ☐ T   ☐ F _____
   _____

7. A disadvantage of finishes for a reverse lay is that they must be pulled from the bed by two persons.
   ☐ T   ☐ F _____
   _____

8. Transverse hose beds are generally located to the front of the apparatus and deployed only to one side.
   ☐ T   ☐ F _____
   _____

MATCHING: Write the correct letter in the space provided.

9. Match types of hose loads to their uses.
   ____ Accordion   A. Supply only
   ____ Horseshoe   B. Attack only
   ____ Flat   C. Supply or Attack

10. Match tools and equipment to the type of equipment strip.
    ____ SCBA   A. Full strip only
    ____ Spare SCBA bottles   B. Partial strip only
    ____ Wye or Siamese   C. Full and partial strip
    ____ Ladders
    ____ Pike Poles
    ____ Forcible Entry Tools

SELECT: Circle the correct response.

The "U" portion of the horseshoe in a horseshoe load is at the **11.** (front, rear) of the bed, while the "U" in a reverse horseshoe is at the **12.** (front, rear) of the bed.

SHORT ANSWER: Answer each item briefly.

13. The procedure for laying hose from a main hose bed is the same for both attack and supply hose, except that there is one additional step when laying attack hose. What is this step?

    _____

    _____

14. What is a hose load finish?

    _____

    _____

15. What are the two categories of finishes?

    _____

    _____

16. Which category of finish is the more elaborate?

    _____

    _____

17. Which hose appliance(s), other than a nozzle, is/are used for the cisco finish?

    _____

    _____

18. Which hose appliance, other than a nozzle, is used for the reverse horseshoe finish?

    _____

    _____

218 HOSE

19. Which hose appliance, other than a nozzle, is used for the skid load finish?

20. What are some of the advantages and disadvantages of carrying preconnected attack hose in longitudinal hose beds or raised trays?

21. Hose loaded in tailboard compartments has many of the same advantages and disadvantages as hose loaded in longitudinal beds, but also has other disadvantages. What are these additional disadvantages?

22. What are some of the advantages and disadvantages of carrying preconnected attack hose in transverse hose beds?

23. Hose loaded in side compartments or bins has many of the same advantages and disadvantages as hose loaded in transverse hose beds, but also has other disadvantages. What are these additional disadvantages?

24. Besides the fact that it is preconnected, what is the major difference between a flat load used for supply hose and flat load used for preconnected attack hose?

25. What is the major advantage and disadvantage of a minuteman load?

LISTING

26. List the three ways in which attack hose can be arranged when it will be used for attack only, not as supply hose.

   A. _____
   B. _____
   C. _____

Attack Hose Loads and Layout Procedures **219**

27. List the three finishes for a reverse lay.

    A. _____

    B. _____

    C. _____

28. List the two major advantages of carrying hose that is preconnected to a pump discharge valve outside the main hose bed.

    A. _____

    B. _____

29. List the seven locations where preconnected attack lines can be carried on an apparatus.

    A. _____

    B. _____

    C. _____

    D. _____

    E. _____

    F. _____

    G. _____

30. List the five types of loads described in this manual for preconnected attack hose.

    A. _____

    B. _____

    C. _____

    D. _____

    E. _____

31. List the two primary uses for hose packs.

    A. _____

    B. _____

32. List the three ways in which a hose pack may be carried.

    A. _____

    B. _____

    C. _____

DISCUSSION QUESTIONS

What changes would you recommend regarding the way attack hose is presently carried on your apparatus? Why do you feel these changes are warranted?

_____

_____

_____

220 HOSE

Which type of finish do you feel is best for a reverse lay? Why?

_____
_____
_____

What changes would you recommend in your department's procedures for standpipe packs? Why do you feel these changes are warranted?

_____
_____
_____

7

Special Hose Operations

222 HOSE

This chapter provides information that addresses performance objectives described in NFPA 1001, *Fire Fighter Professional Qualifications* (1987), particularly those referenced in the following sections:

Fire Fighter I

3-12 Ladders.

3-12.7

3-13 Fire Hose, Nozzles, and Appliances.

3-13.2

3-13.3

3-13.4

3-13.11

Fire Fighter II

4-13 Fire Hose, Nozzles, and Appliances.

4-13.5

Chapter 7
Special Hose Operations

This chapter contains a number of varied and unrelated hose-handling methods. These special operations are not encountered routinely, but nevertheless, are important to know.

CONNECTING HARD SUCTION HOSE

When using hard suction hose for drafting, it is necessary that the couplings be tight and that the connection to the pumper intake be tight. A rubber mallet can be used to tighten the connection. Use at least two persons to lift the hose into alignment with the pump intake. With the threads of the female coupling against the male threads of the intake fitting, turn the swivel clockwise to make the connection (Figure 7.1). Then rap the lugs of the swivel several times with a rubber mallet to complete the watertight connection (Figure 7.2). Place a chafing block under the hose at the point where it contacts the ground to prevent wear to the hose from pump vibration.

Figure 7.2 Rap the lugs of the swivel several times with a rubber mallet to complete the watertight connection.

Figure 7.1 Turn the hard-suction swivel clockwise to make the connection.

CONNECTING SOFT SLEEVE HOSE

Short lengths of soft sleeve LDH are often carried on an apparatus so that they can be reached quickly when connecting to a hydrant. As shown in

Figure 7.3, such hose can be tied or clamped on the exterior of the apparatus such as in a front bumper well, or placed in compartments. Preconnecting one end of the hose to a suction valve speeds the hookup operation. Hydrant tools can be attached to the hose or placed nearby in a bag.

Position the apparatus so that when the hose is connected between the pumper intake and the hydrant, it forms a gentle loop (Figure 7.4). Use a rubber mallet to make all connections watertight.

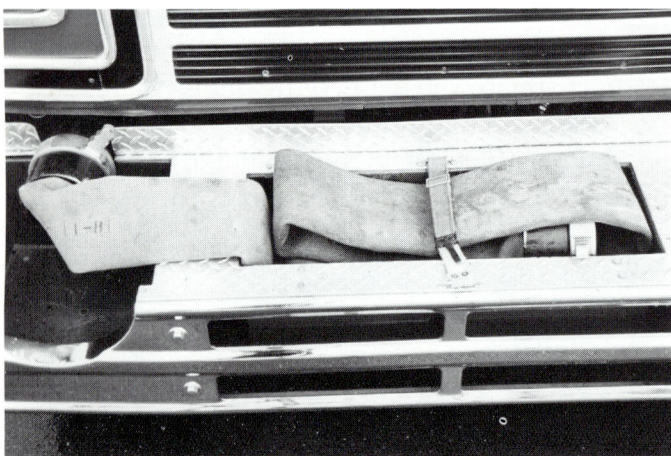

Figure 7.3 The soft sleeve hose can be tied or clamped on the exterior or placed in compartments. This preconnected suction hose is carried in a front bumper well.

Figure 7.4 Position the apparatus so that the suction hose forms a gentle loop.

CONNECTING HOSE TO A PORTABLE MONITOR

Portable monitors are capable of discharging large volumes of water when high-volume nozzles are used. While a portable monitor can be an effective tool when combating an intense fire, it can also become a lethal weapon if not properly secured.

The flow of large volumes of water at 80 to 100 psi (552 kPa to 690 kPa) nozzle pressure produces a significant nozzle reaction. This backward force can quickly tip the monitor backward unless the device is well secured to the ground. A common method for controlling the monitor is to assign at least one person to stay at the device to control the flow and direction of the stream. Occasionally, however, it becomes necessary to leave a monitor unattended, such as when it is used to cool a flammable liquid tank heavily involved in fire. In this case, conditions are too hazardous to leave anyone at the monitor. One of the best ways to secure an unattended portable monitor is to place the hoselines that supply water to the monitor in such a manner that they counteract the nozzle reaction.

When connecting hoselines to any portable monitor, whether it is to be attended or not, lay sufficient hose so that it can be looped *in front* of the monitor. Position the monitor so that the inlets are forward, toward the fire. Lay the hose around to the front in large loops and cross the lines that connect to the outside inlets. Tie the hoselines together with a rope or strap at the crossing point to provide additional stability (Figure 7.5). If using a three-inlet or four-inlet monitor, lay hoses to the inner inlets in the same manner as the outside inlet hoses and secure them to the outside lines as shown in Figure 7.6. (With multiple-inlet monitors, always start connections at the outside inlets and work inward as you connect additional lines.)

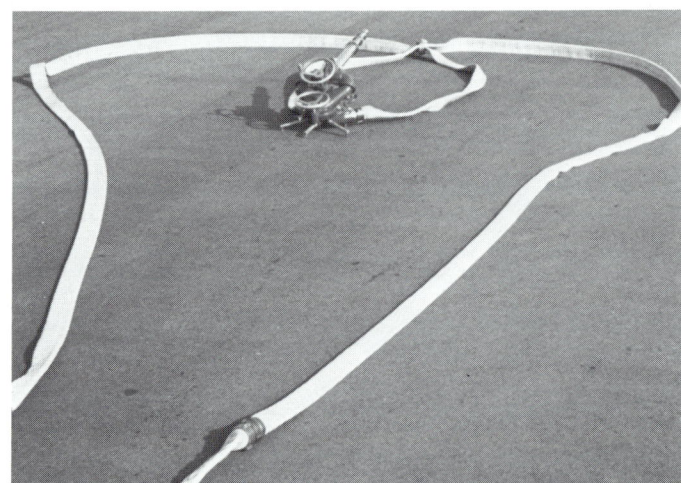

Figure 7.5 With the monitor inlets forward, lay the hose around to the front in large loops and cross the hoses to the outside inlets. Tie the lines together with a rope or strap at the crossing point to provide additional stability.

Special Hose Operations **225**

hoselines were connected directly to the inlets from the rear, it would be necessary to pull the entire hose layout backward to move the monitor.

KINKING HOSE TO SHUT DOWN A CHARGED LINE

When it becomes necessary to shut down a hoseline, as when a section of hose bursts, the simplest method is to close the discharge valve that controls water flow into the hose. Another method is to apply a hose clamp at the coupling closest to the burst section. If a clamp is not available, however, it is sometimes possible to shut down the line by kinking the hose.

To kink a hose, there must be sufficient slack along the length of hose to make a loop, in the same manner as when rolling a loop to advance a nozzle (see Chapter 4, Rolling a Loop to Advance a Charged Hoseline). The steps for kinking a hose are as follows:

Step 1: Starting at a point well ahead of the burst section, straighten the hose progressively, moving from the water source toward the nozzle. As the slack hose accumulates, it will tend to form an "S" shape (Figure 7.8).

Figure 7.6 Lay hoses to the inner inlets in the same manner as the outside inlet hoses and secure them to the outside lines.

Another advantage to arranging the hoselines in this manner is that it is easier to move the monitor backward if fire conditions become untenable (Figure 7.7). If the inlets were to the rear and the

Figure 7.7 Move the monitor backward between the hoses.

Figure 7.8 STEP 1: Straighten the hose progressively from the water source toward the nozzle to form an "S" shape in the hose.

226 HOSE

Step 2: Lay one segment of the "S" over to form a small loop and stand it upright (Figure 7.9).

Step 3: Press the loop downward to form a fold. Kneel on the bends (Figure 7.10). **NOTE:** Water pressure must be released to allow the hose to collapse into a flat fold. This will occur naturally if water is flowing from the burst portion of the hose. If using this method on unruptured hose, however, open the nozzle to release pressure.

RETRIEVING A CHARGED "WILD LINE"

A "wild line" is a hoseline that is uncontrolled at the point of water discharge. Such a line is dangerous because the end of the hose whips from side to side with tremendous force. A wild line condition occurs when a pressurized open nozzle is inadvertently dropped, when a dry hoseline with an open nozzle is charged without anyone at the nozzle, or when a coupling pulls loose from a pressurized hose.

Closing a valve to shut off the flow of water is the safest way to control a wild line. Sometimes this cannot be done quickly enough and direct action must be taken at the uncontrolled end. This is done by lying down on the line some distance behind the flailing end. Maintaining a low profile, crawl forward along the line with one arm fully extended ahead on the hoseline (Figure 7.11). When the open end is reached, pin the end of the hose to the ground and, if a nozzle is attached, close the nozzle (Figure 7.12). If there is no shutoff on the end of the hose, maintain a firm grip on the hose end until the line can be shut down by someone else (by closing a valve or clamping the hose).

Figure 7.9 STEP 2: Lay the "S" over to form a small loop and stand it upright.

Figure 7.11 Crawl forward along the line with one arm fully extended ahead on the hoseline.

Figure 7.10 STEP 3: Press the loop downward to form a fold and kneel on the bends.

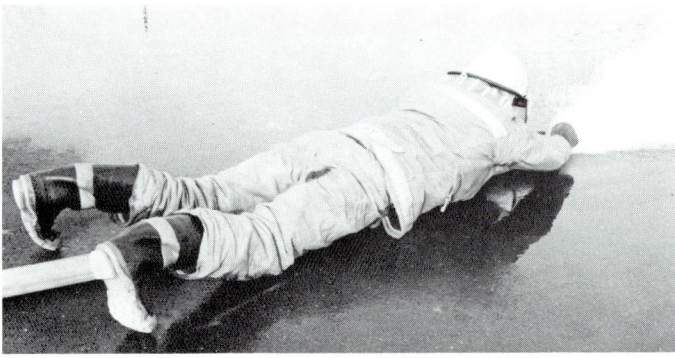

Figure 7.12 Pin the end of the hose to the ground and close the nozzle.

SECURING A HOSELINE TO A GROUND LADDER

To safely direct a fire stream from a ground ladder, it is necessary to secure the hoseline to the ladder. A safe aboveground fire fighting operation requires that the hoseline, the ladder, and the firefighter directing the stream be secured. The firefighter should lock in on the ladder (with a leg lock or a ladder strap) before starting any work above the ground. Tie the ladder down to prevent it from moving as a result of the backward force of the nozzle reaction.

To secure the hose to the ladder, first advance the uncharged line to the appropriate level on the ladder. Then pass the nozzle (which must be *closed*) and approximately 2 feet (0.6 m) of hoseline between two rungs and allow the hose to drape over the bottom rung. At a point two or three rungs below the nozzle, tie the hose to the ladder with a rope hose tool or strap (Figure 7.13).

With the hose secured to the ladder, climb down several rungs, lock in, and call for the line to be charged (Figure 7.14). It is wise to stay clear of the nozzle in case the hoseline is charged too quickly, which could cause the line to straighten suddenly and shift position against the ladder. Standing too close could place you in some danger of being pushed off balance. After the line is charged, readjust the rope or strap, if necessary, to correctly position the hose.

Figure 7.14 Climb down several rungs to stay clear of the nozzle as the hose is charged.

Figure 7.13 At a point two or three rungs below the nozzle, tie the hose to the ladder with a rope hose tool or strap.

Finally, take a position on the ladder to comfortably direct the stream, lock in, and open the nozzle (Figure 7.15).

HOISTING A HOSELINE

Hose can also be hoisted up the outside of a building with a rope. It is considerably easier to hoist an uncharged line than a charged line because it is much lighter, but a charged line can be hoisted if necessary. Use a hose roller where the rope and hose pass over the edge of the roof or windowsill.

To hoist an uncharged hoseline, fold the first 6 feet (1.8 m) of hose back on itself. If a nozzle is attached, fold the hose so that the shutoff handle is against the hose. This will help prevent the handle from snagging as the hose is hoisted and allow the end to pass over the hose roller more easily.

Tie the end of the rope, which has been lowered to the ground from above, with a clove hitch around the doubled hose near the nozzle, then tie a half hitch halfway between the nozzle and the folded end. Finish with a half hitch over the doubled hose about 6 inches (15 mm) from the folded end (Figure 7.16).

Hoist the hose to the roof or window with the rope passing over the hose roller. The person on the ground can assist by holding the hose away from the side of the building (Figure 7.17). This helps prevent the hose from snagging on windowsills or other objects.

When a charged hoseline is being hoisted, the nozzle should point upward. Tie the rope to the hose approximately 6 feet (1.8 m) below the nozzle with a clove hitch. Then tie a half hitch at the base of the nozzle on the hose coupling and finish with a half hitch over the tip (Figure 7.18). If desired, pass the short section of rope between the two half hitches over the nozzle shutoff to secure it in the closed position.

Hoist the hose to the roof or window in the same manner as with uncharged hose, with the

Figure 7.15 Lock in and direct the stream from the ladder.

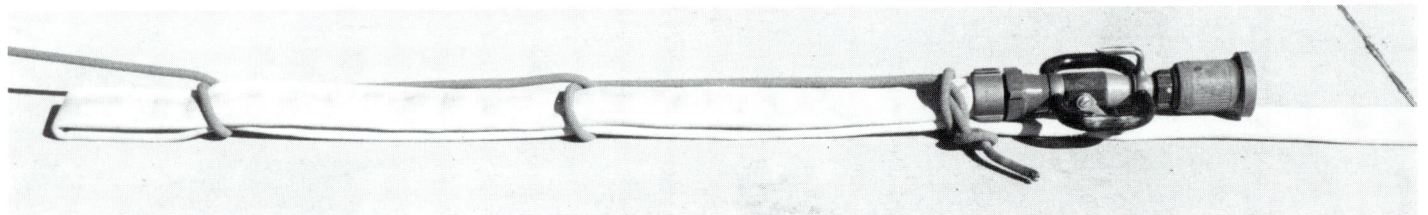

Figure 7.16 Tie a clove hitch around the doubled hose near the nozzle, then a half hitch halfway between the nozzle and the folded end. Finish with a half hitch over the doubled hose about 6 inches (15 mm) from the folded end.

Special Hose Operations **229**

rope passing over the hose roller. Two or more persons may be required to hoist the hose, depending upon the size and total lengths to be hoisted.

When hose hangs on the outside of the building, it should be supported along its length to relieve pressure on the couplings, which must not only withstand internal water pressure, but also the external gravitational force from the hose and water. Place rope or straps on the hose at regular intervals along the hose and secure them to the building at the windows or at the roof line. Another method is to thread the hoseline in and out of windows at every other floor to take the weight off couplings.

NOTE: Hose can be lowered by reversing the procedure for hoisting uncharged hose. It is not recommended that charged hose be lowered to the ground; instead, drain all hose before lowering it to the ground by clamping the hose above a coupling on the ground, disconnecting the hose, and slowly releasing the clamp.

PASSING A HOSELINE UPWARD

If it is not possible to hoist hose with a rope, it can be advanced up the outside of a building from window to window. This operation is performed with a pike pole long enough to reach from one floor to the next. (This can also be done on a fire escape with a firefighter at each landing.)

Start the procedure by folding 3 feet (0.9 m) of hose (with nozzle attached, if desired) back on itself and attaching a hose strap or tool to the fold. Tie the rope in such a way that a loop is formed so that the pike pole hook can slip through the rope (Figure 7.19 on next page).

With a firefighter on each floor at vertically aligned windows, pass the pike pole upward, hook end down and attached to the rope, from firefighter

Figure 7.17 Hoist the hose to the roof or window with the rope passing over the hose roller.

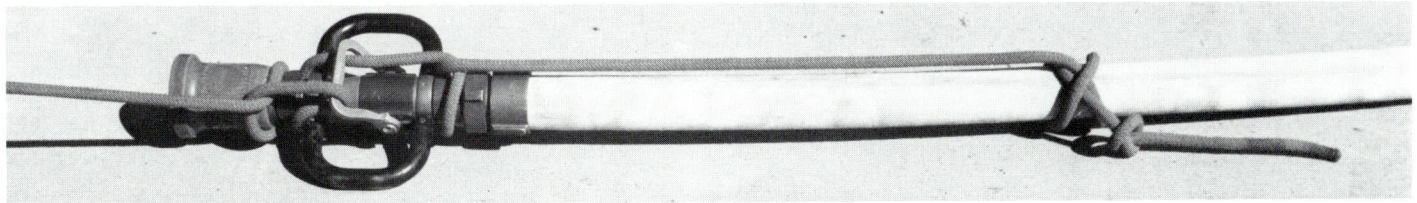

Figure 7.18 Tie a clove hitch 6 feet (1.8 m) below the nozzle, a half hitch at the base of the nozzle, a loop through the nozzle bale, and a half hitch over the tip.

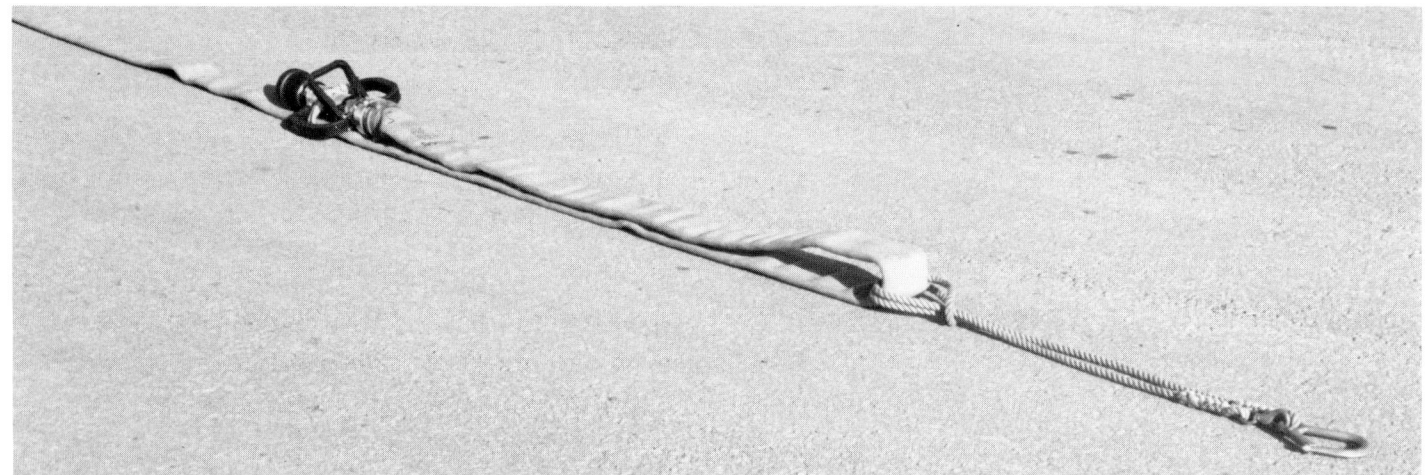

Figure 7.19 Fold 3 feet (0.9 m) of hose back on itself and attach a hose rope tool so that a loop is formed.

to firefighter at each window so that the hose moves progressively upward (Figure 7.20). As the hose ascends to its destination, each person should continue to lift the hose below to relieve pressure on the couplings and to make it easier for the firefighters above to pass the pike pole and hose upward.

When the evolution is completed with the hose in position for fire fighting, secure it to the building at each window with hose straps or ropes.

Figure 7.20 Pass the pike pole and hose upward from firefighter to firefighter.

Chapter 7 Review
Answers on page 238

TRUE-FALSE: Mark each statement true or false. If false, explain why.

1. A rubber mallet should be used to make both hard and soft sleeve hose connections watertight.
   ☐ T ☐ F _____

2. When a soft sleeve LDH is connected between a pumper intake and a hydrant, the hose should be straight.
   ☐ T ☐ F _____

3. Hoselines connected to a portable monitor should be looped in front of the monitor with the monitor inlets forward, toward the fire.
   ☐ T ☐ F _____

4. With multiple-inlet monitors, you should start connections at the outside inlets and work inward as you connect additional lines.
   ☐ T ☐ F _____

5. When directing a fire stream from a ground ladder, the ladder should be tied down.
   ☐ T ☐ F _____

6. When advancing a hoseline up a ground ladder, it is best to charge the hose before climbing the ladder.
   ☐ T ☐ F _____

7. When directing a fire stream from a ground ladder, the hose should be tied to the ladder two or three rungs below the nozzle.
   ☐ T ☐ F _____

8. If it is necessary to hoist a charged hoseline upward, the nozzle should point downward.
   ☐ T ☐ F _____

232 HOSE

9. If a hose cannot be hoisted upward by a rope, it can be advanced up the outside of a building from window to window.
   ☐ T ☐ F _____
   _____

SHORT ANSWER: Answer each item briefly.

10. What is the minimum number of persons that should be used to make a hard suction hose connection?
    _____
    _____

11. If it is not possible to shut off the flow of water to a wild line, what can the firefighter do to close the nozzle?
    _____
    _____

12. A hose which has been hoisted upward should be drained before being lowered to the ground. What is the recommended method for draining the hose?
    _____
    _____

LISTING

13. When a section of hose bursts, list the three recommended methods of shutting down the line. (List the simplest method first.)
    A. _____
    B. _____
    C. _____

14. If hose is hoisted upward on the outside of a building, it should be supported to relieve the pressure on the couplings. List the two methods of accomplishing this.
    A. _____
    B. _____

Review Answers

CHAPTER 1

1. C
2. A
3. D
4. False. Most fire hose is double-jacketed
5. True
6. True
7. False. It is usually carried on reels
8. False. It is made in ¾- and 1-inch (19 mm and 25 mm) sizes
9. True
10. True
11. False. Different fabrics can be used
12. True
13. A
    C
    D
    B
14. C
    E
    B
    D
    A
15. 400 (2 758), 1,250 (8 619)
16. Fighting fires in wildland settings
17. Single-jacketed
18. Either lined or unlined
19. It is more heat resistant
20. Single-jacketed
21. Either lined or unlined
22. 1½-inch (38 mm)
23. It becomes unreliable with age
24. Relay-supply hose
25. Water friction on the inner lining
26. It is better able to withstand vacuum conditions
27. Thermoplastic
28. Thermoplastic
29. Processes for manufacturing rubber hose liners
30. They aid in tightening and loosening connections
31. A. Lay several parallel lines
    B. Place pumpers at intervals to boost pressure
32. A. 3½ inch (90 mm)
    B. 4 inch (100 mm)
    C. 4½ inch (115 mm)
    D. 5 inch (125 mm)
    E. 6 inch (150 mm)
33. A. Forging
    B. Extruding
    C. Casting
34. A. Threaded
    B. Sexless
    C. Snap
35. A. Quarter-turn
    B. Storz
36. A. Expansion ring method
    B. Screw-in expander method
    C. Collar method
    D. Tension ring method
    E. Banding method

CHAPTER 2

1. B
2. C
3. C
4. False. Once a fire is contained, time can be taken
5. False. The glass should be completely removed from the window opening
6. True
7. False. It should be dragged flat
8. True
9. True
10. False. Heat will dry the liner, which promotes cracking
11. True
12. False. Chemicals in gaseous form are also injurious
13. False. Laying hose near curbs can result in chemical damage
14. False. Most damage is caused by rough handling

234 HOSE

15. True
16. True
17. False. Minor damage can be repaired
18. True
19. False. Hose should be secured to prevent damage to couplings
20. True
21. True
22. True
23. False. Hold them above the pump discharge to promote the purging of air
24. False. This is only the pressure at which the hose is charged with water. Test pressures are much higher, depending on the type of hose.
25. False. It should be maintained for five minutes
26. True
27. False. It can be pressure tested
28. True
29. C
    D
    A
    B
30. 10 (254); 10
31. Yes; Improvise padding, such as with a rolled salvage cover
32. A. Move the debris
    B. Cover the debris with canvas
33. To allow a vehicle to drive over a hose without damaging the hose
34. A chafing block
35. Four times
36. A sudden increase in pressure
37. Closing nozzles too quickly
38. Close nozzles slowly and use pressure control devices
39. By allowing some water to flow through the nozzle at all times
40. Wash the hose with a mild soap solution and dry completely
41. Wash the hose with a solution of bicarbonate of soda and water, rinse thoroughly, and dry completely
42. Remove the gasket and wash the coupling in warm, soapy water. Clean the threads with a stiff brush.
43. Once a year
44. Yes
45. No
46. Yes
47. Yes
48. No (synthetic hose)
49. Yes
50. Yes
51. A. By mechanical action of objects
    B. By exposure to heat or cold
    C. By mold or mildew
    D. By chemical contact
52. A. Melt the ice with a steam-generating device
    B. Carefully chop away the ice to free the hose
53. A. Dry hose completely before loading or storing
    B. Cover hose beds with water repellent covers
    C. Ventilate all areas where hose is kept
54. A. Manually operated
    B. Hand-hydraulic
    C. Power
55. A. To check for leaks
    B. To check for a loose lining

CHAPTER 3

1. A
2. A
3. D
4. True
5. False. By the definition of "appliance," couplings are not considered appliances
6. False. Fog nozzles are used on master stream devices
7. False. Water curtains do not stop radiant heat
8. True
9. False. It is not recommended because the valve and handle can suddenly snap open, possibly causing injury to the pump operator
10. True
11. True
12. True
13. False. They can be connected with a double male adapter
14. True
15. True

16. E
    D
    C
    A
    B
17. A
    C
    B
    D
18. 50 (345)
19. 80 (552)
20. 100 (690)
21. 20 (6), 5 (2)
22. Shutoff valve
23. Tip
24. Turret pipe or deck gun
25. Ladder pipe
26. Ball valve
27. Clapper valve or check valve
28. It permits a pumper to increase pressure on a supply hose
29. It permits a hydrant to be turned on without an operator being present
30. In a water thief, one discharge orifice is as large as the intake orifice; in a manifold, all discharge orifices are smaller than the intake orifice
31. Wye
32. The inlet has a female fitting; the outlets are male fittings
33. Siamese
34. The inlet has female fittings; the outlet is a male fitting
35. It prevents pump damage by diverting a sudden pressure surge away from the pump
36. Plug
37. Cap
38. Blindcap
39. By using reducers or increasers
40. Elbow
41. Silicone or graphite
42. Hose control device
43. It can be placed on a hose to contain a leak
44. It allows vehicles to pass over a hose without damaging the hose
45. A. 2½-inch (65 mm)
    B. 3-inch (77 mm)
46. A. Gate
    B. Ball
    C. Butterfly
    D. Floating
    E. Clapper
    F. Piston

CHAPTER 4

1. B
2. D
3. C
4. True
5. True
6. True
7. False. It is best suited for 1½-inch (38 mm) hose
8. True
9. True
10. False. You should first drape the hose over one shoulder
11. True
12. False. The nozzle should hang at the back
13. True
14. False. They should be used by all persons operating a line
15. False. They should be staggered on opposite sides
16. Female
17. Male
18. Close to
19. Female
20. Counterclockwise
21. Clockwise
22. Counterclockwise
23. Nine
24. 4 (1); 10 (3)
25. Anyone not carrying spanners would be unable to break a spanner-tight connection
26. The gasket is damaged or missing
27. The coupling tilt and between-the-feet methods
28. Holding the other swivel so that it will not turn
29. Female
30. Both couplings are on the outside of the donut roll; only one coupling is on the outside of the straight roll
31. Carrying, because the hose is less subject to abrasion

32. One hand behind the nozzle and the other on the shutoff valve
33. One hand behind the nozzle and the other on the shutoff valve
34. A. Coupling tilt method
    B. Between-the-feet method
    C. Across-the-leg method
35. A. Knee-press method
    B. Stiff-arm method
    C. Spanner wrench method
36. A. Tie the hose to a fixed object such as a pole
    B. Loop the hose over itself and tie it at the crossover point

CHAPTER 5

1. True
2. False. It can also deliver small amounts of water with great efficiency
3. False. The pumper must go to the water source
4. True
5. False. Overpacking may cause couplings to hang up
6. True
7. False. The firefighter should not pull hose until given an order to do so
8. True
9. Supply
10. Attack
11. Maximum
12. Female
13. Male
14. 1,500 (450 mm)
15. A. 200-250 (757-946)
    B. 300-350 (1 136-1 325)
    C. 750-1,000 (2 839-3 785)
    D. 1,200-1,500 (4 542-5 678)
16. 5
17. 750 feet (229 m)
18. Forward lay
19. Male
20. Female
21. Reverse lay
22. Male
23. Female
24. Reverse
25. Split or combination lay
26. Flat and reel
27. Straddle the hose with a pumper and drive slowly forward as the hose is loaded into the bed
28. Reverse lay
29. A. The mission of the apparatus
    B. The size of the apparatus pump
    C. The nature of the water source
    D. Whether or not the unit will work independently
30. A. Forward lay
    B. Reverse lay
    C. Split or combination lay
31. A. Accordion
    B. Horseshoe
    C. Flat
32. Step 1: Place the hydrant tools near the hydrant
    Step 2: Pull enough hose to wrap around the hydrant
    Step 3: Wrap the hose around the hydrant and signal the driver to lay the hose
    Step 4: Unwrap the hose and bring the end to the appropriate outlet
    Step 5: Remove the outlet cap and place the wrench on the valve stem
    Step 6: Connect the hose to the outlet
    Step 7: Open the hydrant valve at the appropriate time
33. A. When knockdown can be achieved with the tank water
    B. When the supply hose can be charged within one or two minutes

CHAPTER 6

1. C
2. B
3. True
4. True
5. True
6. False. The male coupling is on the outside
7. False. One person can pull the finish from the bed
8. False. They deploy to either side
9. C
   C
   C
10. C
    A
    C
    A
    A
    C
11. Front
12. Rear
13. Enough additional hose must be pulled to reach the rear of the structure
14. An arrangement of hose placed on the top of a hose load
15. Those for forward lays and those for reverse lays
16. Those for reverse lays
17. A wye
18. A wye
19. None
20. Advantage: The hose is in an ideal position when the pumper stops past a fire or when a reverse lay is made
    Disadvantages: The pump operator cannot see which hose is pulled from the rear of the apparatus; the hose may become entangled on the apparatus if led to the side
21. The amount of hose that can be loaded is limited and loading can be difficult
22. Advantages: The hose is close to seated firefighters and separated from supply hose, the pump operator can see the hose as it is pulled, and the hose will not become entangled on the apparatus when pulled to the side
    Disadvantages: The hose is more distant when a pumper stops past a fire
23. There is a delay if the fire is on the opposite side of the apparatus; space is limited
24. Loops are provided on the preconnected attack hose to expedite the pulling of hose
25. Advantage: It is carried clear of the ground so it will not snag
    Disadvantage: It is awkward to carry when wearing an SCBA
26. A. In a finish load attached to the end of a hose bed load
    B. In a preconnected, quick-attack load
    C. In a portable hose pack
27. A. Cisco
    B. Reverse horseshoe
    C. Skid
28. A. Time is saved when making a quick attack
    B. There is no need to move it when laying supply hose
29. A. Longitudinal beds
    B. Raised trays
    C. Transverse beds
    D. Tailboard compartments
    E. Side compartments or bins
    F. Front bumper wells
    G. Reels
30. A. Flat
    B. Minuteman
    C. Reverse horseshoe
    D. Triple layer
    E. Wyed flat
31. A. Connecting to standpipes in high-rise buildings
    B. Laying out hose in wildland fire fighting situations
32. A. On a firefighter's back
    B. By hand
    C. Secured to a cart

CHAPTER 7

1. True
2. False. It should form a gentle loop
3. True
4. True
5. True
6. False. The hose should be uncharged for ease of maneuverability
7. True
8. False. It should point upward
9. True
10. Two
11. Lay down on the line, then crawl forward toward the nozzle with one arm fully extended ahead on the hoseline, pin the open end to the ground when it is reached, and close the nozzle
12. Clamp the hose above a coupling on the ground, disconnect the hose, then slowly release the clamp
13. A. Shut off water flow to the hose
    B. Apply a hose clamp at the closest coupling
    C. Kink the hose
14. A. Tie the hose to the building at regular intervals with ropes or straps
    B. Thread the hoseline in and out of windows at every other floor

Index

A
Acceptance testing, x, 48
Accordion load, x, 148-151
   advantages and disadvantages of, 148-149
Accordion shoulder carry, 148, 152
   for advancing hose up stairway, 127
   from hose in accordion-loaded hose bed, 112-114
   from hose on ground, 111-112
   from hose in other hose beds, 114-115
   laying out, 115
Adapter, x, 77
Adjustable flow nozzle, x, 67
Advancing a booster line, 129-135
   one-person methods, 131-132
   three-person methods, 134
   two-person methods, 132-134
Advancing a charged hoseline
   rolling a loop to, 135
Advancing hose up a ladder, 128-129
Advancing hose up a stairway, 127-128
Advancing hoselines, 125-135
   arm-lock method, 126
   hose-strap method, 126
Aluminum alloy couplings
   advantage of, 17
   disadvantage of, 17
American Insurance Association, 6
American National Fire Hose Connection
   Screw Thread. *See* National Standard Thread
American National Standards Institute (ANSI), 6
American Society for Testing and Materials (ASTM), 6
American Water Works Association, 6
Anchoring and connecting hose to hydrant, 161-162
Apparatus pump
   size of and hose loads, 143
Appliance(s), x
   Care and maintenance of, 79
Appliances and tools, 67-89
Applicator pipe, x, 69
Association of Factory Mutual Insurance Companies, 6
Attack, initial, 143

Attack hose
   advantage of carrying outside main hose bed, 189
   defined, x, 7
   disadvantage of pumper at water source, 175
   laying from supply hose bed, 175-179
   loads and layout procedures, 175-220
   single in forward lay, 176-178
   single in reverse lay, 178-179
   and supply hose on two separate apparatus, 145
Automatic hydrant valve(s), x
   mechanically controlled, 74-75
   radio controlled, 74-75

B
Ball valve, x, 71
Banding method, x, 22
Bar, straight, 133
Battery acid
   and hose damage, 35
Blindcap(s), x, 78
Blunt start, x, 18
Bolted-on collars
   couplings attached with, 46
Booster hose, x
   construction of, 16
   lugs for, 18
Bowl, 18
Braided hose, x, 16
Bucket brigade, 1
Buddy system, 126
Butterfly valve, x, 71-72

C
Calendering, x, 13
Cap(s), xii, 78
Care and maintenance of hose, 37-40
Carry(ies)
   accordion shoulder, 111-115
   accordion shoulder, modified, 115-117
   horseshoe shoulder, 117-118
   horseshoe underarm, 118
   shoulder loop, 108-111
Carts, hose, 40

Cast coupling, x, 17
Chafing block, 33, 223
Chain(s), 80
Chain hose tool, x, 80
Charged hoseline
    kinking hose to shut down, 225-226
    minimum personnel for, 126
Charged wild line
    retrieving, 226
Chemical damage, 35-36
Chlorosulfonated polyethylene, 15
Clamp(s), hose, xii, 81-82
Clapper valve, x, 72, 76
Cold damage, 34
Collar method, x, 22
Combination lay. *See* Glossary
Communication, 161
Connection(s)
    across-the-leg method, 95
    basic rules for, 93-94
    between-the-feet method, 95
    coupling tilt method, 94
    hose to fixed fitting(s), 98
    making and breaking of, 93-98
    nozzle to hose, 98
    one-person methods, 94-95
    two-person methods, 95-96
Contractual sleeve coupling (gas coupling), x
Cotton-polyester jackets
    advantage of, 15
Coupling(s), x
    aluminum alloy, 17
    attached with bolted-on collars, 46
    attached with screw-in expanders, 44
    attached with tension rings, 47-48
    attaching to hose, 21-22
    banding, 22
    blowing, 17
    cast, 17
    collar, 22
    construction of, 5-27
    contractual sleeve, x
    expansion ring, 21
    with expansion rings, 41-44
    extruded, xi, 17
    forged, xi, 17
    repair of, 36
    screw-in expander, 21-22
    sexless, 19-20
    snap, 20-21
    standards, 5-7
    tension ring, 22
    threaded, 18-19
    three-piece, 19
    types of, 17-21
Coupling damage
    causes and prevention of, 36-37
Coupling thread
    female, 18
    male, 18
Crimp, 15
Curing, xi, 15

D

Damage to hose
    causes and prevention of, 31-36
Debris
    and hose damage, 32
Deck gun, xi. *See also* Turret pipe
Deluge set. *See* Monitor(s)
Donut roll, xi
    one-person, 101-103
    twin, 104-105
    twin, self-locking, 106-108
    two-person, 103-104
Drag(s)
    multiple sections, 122-123
    single section, 119-121
Dragging
    hose abrasion from, 32
Drop-forged coupling, xi. *See also* Forged coupling
Dryers, hose, xii, 39
Drying hose, 38-39
Dutchman, xi, 47

E

Eductor, xi, 78
Elbow(s), 77
Equipment strip, xi, 175
Expander(s), xi
    types of, 41
Expansion ring(s), xi, 41-44
    procedure for attaching female, 42
Expansion ring gaskets, 42
Expansion ring method, xi
Extinguisher hose, xi, 10, 16

Extruded coupling, xi, 17
Extrusion, xi, 17

F
Factory Mutual (FM), 6
Female coupling, xi, 18
Filler yarn, xi, 14. *See also* Weft yarn
Finish(es), xi, 180-189
   cisco, 181-182
   for forward lays, 180
   reverse horseshoe, 184-186
   reverse lay with cisco finish, 183
   reverse lay with reverse horseshoe finish, 186
   reverse lay with skid load finish, 188-189
   for reverse lays, 181-186
   skid load, 187-189
Fire extinguisher hose
   construction of, 16
   conventional, 10
   high-pressure, 10
Fire service hose, xi
   methods of construction, 11-16
Fitting(s), 67, 77-78
Flat load, xi, 154-157, 198
Flexible hose, 1
Floating valve, xi, 72
Flowmeter, 53
Forestry hose, xi, 7
Forged coupling, xi, 17
Forward Lay, xi, 162-164
Four-way hydrant valve, xi, 165
Front
   meaning of, 149
Front bumper well, xi, 191

G
Gasoline
   and hose damage, 35
Gate valve(s), xi
   nonrising stem valve, 71
Gated wye, xii, 71, 76
Governor, 34

H
Handles, 18-19
Hand-powered pump, 1

Hard suction hose, xii, 10
   connecting, 223
   vacuum testing of, 53-55
Heat damage, 34
Higbee cut, xii, 18
Higbee indicator, 18, 36
Horseshoe load, xii, 152-154
   advantages and disadvantages of, 152
Hose(s)
   appliances and tools, 67-89
   attack, 7-8, 175-179
   booster, 8, 16
   braided construction, 16
   care and maintenance of, 37-40
   carries and drags, 108-125
   carts, 40
   chemical treatment of, 15
   clamps, 81-82
   classification by construction, 10-11
   classification by use, 7-10
   connections, making and breaking of, 93-98
   construction, 5-27
   control device, 80
   curing, xi
   drags, 119-125
   drying of, 38-39
   fire, standards, 5-7
   fire extinguisher, 10, 16
   forestry, 7
   handling, basic methods of, 93-139
   hard suction, 10
   hard suction, vacuum testing of, 53-55
   hydrostatic testing of, 15
   industrial, xii, 5
   inspection of, 40-48
   intake, 10
   jackets, 14-15, 81
   large diameter (LDH), 9
   lined, 8
   mill, xiii, 5
   multiple-jacketed, 7
   patching of, 55
   racks, xii, 40
   recoupling, 40-48
   reinforced, rubber-covered, rubber lined, 16
   roller(s), 80-81
   rolling, 15
   rolls, 99-108

242   HOSE

round curing of, 15
rubber-covered construction, 7, 15-16
rubberized, 5
service testing of, 48-55
single-jacketed, 7
small, 194
soft sleeve, 10
standpipe, xiv, 8, 12
storage of, 40
supply, 9
tests and inspections, records of, xii, 55-56
through-the-weave construction of, 15
unlined, xiv, 8
washing of, 37-38
woven fabric-jacket, 7
woven-jacket construction , 12-15
woven-jacket type, 5
wrapped construction, xv, 16
Hose bed(s), xii
  bumper wells, 191
  longitudinal, 190
  reels, 191
  side compartments, 191
  tailboard compartment, 190-191
  transverse, 190
Hose bin, xii, 191
Hose bridge(s), xii, 33, 36, 82
Hose cabinet, xii, 8
Hose cap, xii, 78
Hose carry, xii, 108-118
Hose clamp, xii, 81-82
Hose control device, xii, 80
  Hose damage. *See* Damage
Hose dryer, xii, 39
Hose frozen in ice
  method of removal, 34
Hose jacket, xii. *See also* Jacket(s)
Hose lay procedures, 159-166
  basic guidelines, 159-160
Hose laying operations, 1
Hose lays
  direction of, 145-157
Hose load finishes, 180-189
Hose loads
  basic, 148-157
  determining, 143-145
  guidelines, 147-148
Hose operations, special, 223-232

Hose pack(s), xii, 207-215
  high-rise, 8
  standpipe, 207-209
  wildland, 209-212
Hose rack, xii, 40
Hose ramp(s). *See* Hose bridge(s)
Hose record, xii, 55-56
Hose reel, xii
Hose roller(s), xii, 31, 228-229
Hose strap(s), 80, 126, 129, 131, 133, 229-230
Hose test gate valve, xii, 49
Host tool, xii, 80, 223-224
Hose tower, xii, 38-39
Hose washing machine, 37
Hose wringer(s), xii, 83
Hoseline
  hoisting, 228-229
  passing upward, 229-230
  to ground ladder, securing, 227-228
Hydrant(s)
  LDH and selection of, 160-161
  making a, 160
  yard, 8
Hydrant wrench(es), xii, 79-80
Hydraulic press clamp, 82

I

Increaser coupling, xii, 77
Industrial hose, xii, 5
In-line eductor, xii, 78
In-line relay valve, xii, 76
Intake hose, xiii
  hard suction, 10
  soft sleeve, 10
Intake relief valve, xiii, 76-77
International Association of Fire Chiefs, 6
International Association of Fire Engineers. *See*
  International Association of Fire Chiefs

J

Jacket(s), 14-15, 81
  assembly of, 15
Jet-spray washer, 37

K

Kinking hose
  to shut down charged line, xiii, 225-226

L

Ladder pipe(s), xiii, 69-70
   smoothbore tip and, 68
Ladder strap(s), 227
Large diameter hose (LDH), xiii, 49
   advantages of, 145
LDH discharge valve connection, 49
LDH, piston valves and, 73
LDH loading, 157
LDH and manifold(s), 75
LDH supply lay, 163-165
Leg lock, 227
Lined hose, xiii
   wide use of, 12
Liner(s)
   assembly of, 15
   characteristics of, 12
   manufacture of, 13
Longitudinal hose bed, xiii
Lugs, 18
   types of, 18

M

Making a hydrant, xiii
Male coupling, xiii, 18
Manifold(s), xiii, 75
Master stream, xiii
Master stream devices, 69-70
   smoothbore tip and, 68
Mildew damage, 15, 34-35
Mill hose, xiii, 5
Minuteman load, 202-203
   advantage and disadvantage of, 202
   pulling and advancing, 203
Mold damage, 34-35
Monitor, portable
   connecting hose to, 224-225
Monitor(s), xiii, 69, 224
   smoothbore tip and, 68
Motor oil
   and hose damage, 35
Multiple jacket hose, xiii, 15

N

National Fire Protection Association (NFPA), 5-6, 31
National Standard Thread (NST), xiii, 6
NFPA 1901, *Standard for Automotive Fire Apparatus*, 145

Nitrile rubber compound, 15
Nozzle(s), xiii, 67-69
   adjustable flow, v, 67
   applicator, 69
   attaching to hose, 98
   exposure, 68-69
   fog, 68
   mystery, 68
   piercing applicator, xiii, 69
   reaction, 131
   shutoff valve for, 67
   smoothbore tip for, 67
   solid stream, xiv, 67-68

P

Patching hose, 55
Piercing applicator nozzle, xiii, 69
Pike pole, 229
Pin lugs, 18
Piston valve, xiii, 73
Playpipe. *See* Stream straightener
Plug(s), 78
Polyester jackets
   advantages and disadvantages of, 14
Polyurethane, 13
Portable hydrant. *See* Manifold(s)
Preconnect, xiii
Preconnected hose loads, 189-206
Preconnecting to pump discharge valve
   advantage of, 189
Press-down clamp, 82
Proportioner(s), xiii, 78
Pull-up clamp, 82
Pump(s)
   hand-powered, 1

Q

Quarter-turn coupling(s), xiii, 20

R

Racks, hose, 40
Rear
   meaning of, 149
Recessed lugs, 18
Records
   of hose tests and inspections, xii, 55-56
Reducer coupling, xiii, 77
Reducing wye, xiii

Reel load, xiii, 157-159
    advantage and disadvantage of, 157
Reel loading
    of LDH, 158
Relay-supply hose. *See* Supply hose
Relief valve, 34
Reverse horseshoe loads, 191-197
Reverse lay, xiv, 165-166
Rocker lugs, 18
Roll, straight
    butterfly method of unrolling, 100-101
Roll(s), hose, 99-108
    donut, 101-108
    straight, 99-100
Rope(s), 80, 230
Rope tool, 126
Rubber compounds, 13
Rubber mallet, 223-224
Rubber Manufacturers Association (RMA), 6

S

Screw-down clamp, 82
Screw-in expander(s), xiv, 44
Screw-thread connection, breaking, 96-98
    knee-press method, 96
    spanner wrench method, 97-98
    stiff arm method, 96-97
SDH. *See* Small diameter hose
Service testing of hose, xiv, 48-55
    annual text, 48
    equipment needed, 49
    records of, 55-56
    safety, 49
    test procedure, 50-52
    test site preparation, 48-49
    unlined linen hose, 52-53
Set, 19
Sexless coupling(s), xiv
    adapter and, 77
    advantages and disadvantages of, 19
    connecting, 96
    quarter-turn, 20
    Storz, 20
Shank, 18
Shell, 18
Shoulder loop(s)
    laying out hose, 111
Shutoff nozzle, 50
Shutoff valve, 67

Siamese(s), xiv, 76
Single-jacket hose, xiv
Small diameter hose (SDH), xiv
    advantages of carrying only, 145
Smoothbore tip, 67
Snap coupling(s), xiv, 20-21
    adapter and, 77
    advantage of, 21
    disadvantage of, 21
Soft sleeve hose, xiv, 10
Solid stream nozzles, xiv, 67-68
Spanner(s), 18, 79
Split hose bed, 166
Split lay, xiv, 166
Standard for Screw Threads and Gaskets for Fire Hose Connections, 6
Standards
    hose and coupling, 6-7
Standpipe hose, xiv, 8, 12
Standpipe pack, 207-209
Storage, 40
    inside, 34
Storz coupling(s), xiv, 20
    connecting, 96
Strainer(s), suction hose, 82
Stream straightener, 69
Strip
    full, 175
    partial, 175
Suction hose strainers, 82
Supply hose, xiv
    and attack hose on two separate apparatus, 145
    basic hose lays for, 145-147
    charging of, 161
    forward lay, 145-146
    loads and layout procedures, 143-171
    reverse lay, 146
    split lay, 146-147
    water delivered by, 162
Swivel, 18

T

Tail piece, 18
Tension ring(s), xiv
    couplings attached with, 47-48
Testing of hose
    acceptance testing, 48
    service testing, 48

Thermoplastics, xiv, 13
Thread, standardized, 6
Threaded coupling, xiv
Threads per inch (millimeters), table of, 6
Three-ply process, xiv, 16
Through-the-weave construction, 15
Tier, xiv, 149
Tool(s), 18, 79-80, 126, 229
    rubber mallet, 223-224
    and appliances, 67-89
    or equipment, hose damage from, 32
Transverse hose bed, xiv
Triple layer load, 200-202
    disadvantage of, 200
    pulling and advancing, 202
Tubes. *See* Liners
Turret pipe(s), xiv, 69-70
Twisting, 15

U

Underwriters Laboratories (UL), 6
United States Forest Service (USFS), 6
Unlined fire hose
    forestry, 12
    standpipe, xiv, 8, 12
Unlined hose, xiv, 8

V

Vacuum testing of hard suction hose
    equipment needed for, 53
    procedure for, 53-55
    reasons for, 53-55
Valve(s), xiv, 67, 71-73
    automatic hydrant, 74-75
    ball, 71
    butterfly, 71-72
    clapper, 72, 76
    floating, xi, 72
    hose test gate, xiii, 49
    hydrant booster, xii, 73
    in-line, xii, 76
    intake relief, xiii, 76-77
    LDH discharge connection, 49
    piston, xiii, 73
    relief, 34
    shutoff, 67
Valve devices, 73-78
Venturi principle, 78

W

Wagon, 157
Warp yarn, xiv, 14
Washing hose, 37-38
Water curtain, xv, 68
Water hammer, xv, 34, 76
Water supply unit, 143
Water thief, xv, 75
Waterway, xv, 21-22
Weeping, xv
Weft yarn, xv, 14
Whip method, 131
Wild line, xv, 226
Wildland pack, 209-212
    cords for, 209-210
    progressive hose lay from, 213-215
Window openings
    and hose damage, 32
Woven-jacket hose, xv
    lined, 12
    unlined, 12
Wrapped hose, xv, 16
Wye(s), xv, 76
    gated, xii, 71, 76
Wyed flat load, 204-205

Y

Yard hydrant, 8
Yarns(s)
    filler, 14
    warp, 14
    weft, 14

NOTES

NOTES

NOTES

NOTES

NOTES

HOSE PRACTICES
Seventh Edtiion

COMMENT SHEET

DATE _____ NAME _____
ADDRESS _____
ORGANIZATION REPRESENTED _____
CHAPTER TITLE _____ NUMBER _____
SECTION/PARAGRAPH/FIGURE _____ PAGE _____

1. Proposal (include proposed wording or identification of wording to be deleted), OR PROPOSED FIGURE:

2. Statement of Problem and Substantiation for Proposal:

RETURN TO: IFSTA Editor
Fire Protection Publications
Oklahoma State University
930 N. Willis
Stillwater, OK 74078-8045

SIGNATURE _____

Use this sheet to make any suggestions, recommendations, or comments. We need your input to make the manuals as up to date as possible. Your help is appreciated. Use additional pages if necessary.

Your Training Connection.....

The International Fire Service Training Association

We have a free catalog describing hundreds of fire and emergency service training materials available from a convenient single source: the International Fire Service Training Association (IFSTA).

Choose from products including IFSTA manuals, IFSTA study guides, IFSTA curriculum packages, Fire Protection Publications manuals, books from other publishers, software, videos, NFPA standards, and a free subscription to IFSTA's *Speaking of Fire*.

Contact us by phone, fax, U.S. mail, e-mail, internet web page, or personal visit.

Phone
1-800-654-4055

Fax:
405-744-8204

U.S. mail
IFSTA, Fire Protection Publications
Oklahoma State University
930 North Willis
Stillwater, OK 74078-8045

E-mail
editors@ifstafpp.okstate.edu

Internet web page
www.ifsta.org

Personal visit
Call if you need directions!